Time and Petri Nets

Louchka Popova-Zeugmann

Time and Petri Nets

 Springer

Louchka Popova-Zeugmann
Institut für Informatik
Humboldt-Universität zu Berlin
Berlin, Germany

ISBN 978-3-642-41114-4   ISBN 978-3-642-41115-1 (eBook)
DOI 10.1007/978-3-642-41115-1
Springer Heidelberg New York Dordrecht London

Springer is part of Springer Science+Business Media (www.springer.com)

Preface

Time and Petri nets – do they not contradict each other? While time determines the occurrences of events in a system, classic Petri nets consider their causal relationships and represent events as a concurrent system. At first, these two appear to be at odds with each other, but taking a closer look at how time and causality are intertwined, one realizes that time actually enriches Petri nets. There are many possible ways in which time and Petri nets interact, this book takes a deeper look at three time-dependent Petri nets: Time Petri nets, Timed Petri nets, and Petri nets with time-windows.

The aim of this book is to introduce different algorithms that can be used to analyze these three time-dependent Petri nets, as well as the principal methods for analyzing nets in general. To give the reader a general understanding of Petri nets and their origins, we will first take a look at classic Petri nets and their fundamental properties. Once the basis has been laid, we will take a dive into time-dependent Petri nets, which are an extension of classic Petri nets.

There are many different possibilities to associate time to classic Petri nets. For the three time-dependent Petri nets this book focuses on, time is associated to transitions or to places. For the first nets that we will take a look at, Time Petri nets, enabled transitions may fire only during specified time intervals. The transitions must fire the latest at the end of their intervals if they are still enabled then. At any given moment only one transition may fire. This firing does not take time. For the second class of nets, Timed Petri nets, a maximal set of just-enabled transitions fires, and the firing of each transition takes a specific amount of time. The third class of nets, Petri nets with time-windows, portrays time as a minimum and maximum retention for tokens on places. In these nets tokens can be used for firing only during their minimum and maximum retention. At the end of the maximum retention time for a token its time is reset to zero if it was not used for firing. The next period of its retention time on this place then restarts. This repetition can continue indefinitely.

The pivotal contribution of this book is the introduction of algorithms that allow the analysis of the different kinds of time-dependent Petri nets. For each class of time-dependent nets, we will consider different algorithms that have specifically been invented for the analysis. For Time Petri nets, we provide an algorithm which proves the behavioral equivalence of a net where time is designed once with real and once with natural numbers. One can also say that the *dense semantics* of Time Petri nets can be replaced with *discrete semantics*. The added value of this approach is that at this point we can reduce the state space of a Time Petri net and consider its integer-states exclusively. The result then allows for a qualitative and quantitative analysis.

As a new approach for Timed Petri nets, we introduce two *time-dependent state equations*. These provide a sufficient condition for the non-reachability of states. We also define a local transformation for these nets into Time Petri nets. Eventually we show possible variations of them.

Last but not least, we prove that Petri nets with time-windows have the ability to realize every transition sequence fired in the net omitting time restrictions. Despite the first experience that time has no influence on the behavior of such nets, we verify that the time can change the liveness behavior of Petri nets with time-windows.

Finally, we choose these three classes of time-dependent Petri nets to show that time alone does not change the power of a Petri net. In fact, time can or cannot be used to force firing. For Time Petri nets and Timed Petri nets we can say that they are Turing-powerful, and thus more powerful than classic Petri nets. The reason for this is that there is a compulsion to fire at some point in time. For Time Petri nets this is at the latest point of an interval, and for Timed Petri nets this is immediately after enabling. In contrast to these two nets, Petri nets with time-windows have no compulsion to fire. Their *expressiveness power* is less than that of Turing-machines.

This book is based on a script I have been using for my advanced lecture on *Time and Petri nets*. To read and understand it, you do not need advanced mathematical knowledge, except for the section on *quantitative evaluation of Time Petri Nets*, where some insight into operational research and graph theory will be useful.

This book would not have been possible without the help and support of many people. Special thanks go to Jan-Thierry Wegener, Matthias Werner and Jörg Bachmann for the countless hours they spent discussing Time and Petri nets with me. I would like to thank Maria Tammik for her devotion and the time she invested in proofreading the English version of this book and giving valuable suggestions as to how to improve the content. For drawing all the graphics in the English version, I thank Eva Sandig and Phillipp Schoppmann. Finally, my heartfelt gratitude goes to Monika Heiner, my students and my colleagues for encouraging me to write this book.

Berlin, July 2013   Louchka Popova-Zeugmann

Contents

Preface   v

1   Introduction   1

Notation   6

2   The Classic Petri Net   7

   2.1   Definitions 7

   2.2   State Space 13

   2.3   PN-Computability 17

   2.4   Basic Properies 20

   2.5   Bibliographical Notes 26

   2.6   Exercises 28

3   Time Petri Nets   31

   3.1   Basic Notions 31

   3.2   State space 37

   3.3   TPN-Computability 38

   3.4   State Space Reduction 45

   3.5   RG for Finite TPN 72

   3.6   RG for Infinite TPN 78

3.7 Qualitative Properties 89

    3.7.1 Reachability 89

    3.7.2 Liveness 94

    3.7.3 T-invariants 117

3.8 Quantitative Evaluation 118

    3.8.1 Unbounded TPN 122

    3.8.2 Bounded TPN 128

3.9 Bibliographical Notes 133

3.10 Exercises 135

4 Timed Petri Nets   **139**

4.1 Definitions 140

4.2 Timed PN and Counter Machines 146

4.3 Transformation of a Timed PN into a Time PN 148

4.4 State Equations 151

4.5 Variations of Timed PN 165

    4.5.1 Timed PN with Priorities 165

    4.5.2 Timed PN with Variable Durations 168

4.6 Bibliographical Notes 170

4.7 Exercises 171

5 Petri Nets with Time Windows   **173**

5.1 Definitions 174

5.2 Reachability 178

5.3 tw-Petri Nets and Counter Machines 186

5.4 Liveness 188

5.5 Bibliographical Notes 188

5.6 Exercises 190

A **191**

A.1 Appendix on Time Petri Nets 191

Bibliography **199**

Index **207**

Chapter 1

Introduction

Und überall hingen, lagen und standen Uhren.
Da gab es auch Weltzeituhren in Kugelform,
welche die Zeit für jeden Zeitpunkt der Erde anzeigten.
[...]
"Vielleicht", meinte Momo,
"braucht man dazu eben so eine Uhr."
Meister Hora schüttelte lächelnd den Kopf.
"Die Uhr allein würde niemand nützen.
Man muß sie auch lesen können."
Michael Ende, Momo

Clocks were standing or hanging wherever Momo looked
– not only conventional clocks but spherical timepieces
showing what time it was anywhere in the world
[...]
"Perhaps one needs a watch like yours to recognize them by"
said Momo.
Professor Hora smiled and shook his head.
"No, my child, the watch by itself would be no use for anyone.
You have to know how to read it as well."[1]
Michael Ende, Momo

The objective of this book is to bring into accordance the two obviously completely contrary concepts of time and Petri nets. Introduced by C. A. Petri

---

[1] Trans. J. Maxwell Brownjohn (Doubleday & Company Inc., New York, and Penguin Books Ltd., 1984).

in [Pet62], Petri nets can be used to study concurrency in the sense of causal independence, but they do not directly deal with time; time is involved only implicitly through the causal relationships. However, the explicit indication of time is indispensable for a great variety of real problems. Even qualitative studies of strongly time-dependent systems are very inexact if time is included only implicitly through causality. The question arises, whether one should at all try to describe and analyze such systems using Petri nets and whether it is even possible.

The first publications in which time and Petri nets are connected, were already published a good ten years after the introduction of Petri nets. As expected, time attributes were first assigned to transitions (cf. [Mer74], [Ram74], [MF76] etc.). After occasional studies in the 1970s and some articles in the 1980s (cf. [BM83], [Sta87], [PZ89] etc.) an avalanche of new time-dependent extensions of the classic Petri net followed in the 1990s. Times in the form of durations or intervals were assigned to places, tokens, and input and output arcs. Furthermore, various different firing rules were defined: earliest possible or latest possible firing, firing single transitions or firing in maximal-step mode, firing with certain probabilities, etc. These extensions arose from practical considerations. With some of these newly developed Petri nets, the originally central idea of concurrency can be found merely as simultaneity, but their practical significance is immense.

Classic Petri nets, i.e., Petri nets without any explicit indication of time, are excellent means for the depiction, simulation and analysis of systems of the most diverse origin. In the description of systems by means of classic Petri nets, events are mostly modeled as transitions. The pre- and postconditions of the events are represented by places and are considered as fulfilled if those places contain any tokens. Directed arcs connect all places which are preconditions of an event with the transition which models this event. Directed arcs analogously connect every transition with all places which represent postconditions for the event modeled by the transition. It is possible to refine these models as and when required by adding time to different elements of the net.

In a classic Petri net, any possible situation is expressed by the assignment of tokens to places, i.e., by a marking. Obviously this is a *discrete* description, whereas a situation in a time-dependent Petri net is a snapshot of a marking at a particular point of time. This means that in Petri nets with continuous time, the situation depends on a discrete parameter (the marking) as well as a continuous one (time). Therefore time-dependent Petri nets are a *hybrid* means of describing a system.

When defining a time-dependent Petri net the following generally need to be fixed:

- the type of time extension: time interval or duration.

- the manner of time modeling: continuous or discrete.

- the specific type of elements of the net to which the time extension is assigned: places, transitions or arcs.

- the firing rule.

When defining a firing rule, there are time-independent characteristics that need to be determined, such as:

- solution of conflicts.

- concurrency of a transition with itself (also called auto-concurrency or self-concurrency).

as well as the following specifications:

- firing mode: whether the transitions fire in single firing mode or in sets (steps).

- at what time the transitions fire: compulsive firing immediately after being enabled, compulsive firing at the latest possible moment after being enabled, without compulsion to fire, or according to a random distribution.

The monograph [Sta95] describes these fundamental construction principles and introduces basic classes of time-dependent Petri nets. According to these modeling specifications we systematize the basic classes of time-dependent Petri nets in the chronological order of their emergence.

The seminal studies of Merlin ([Mer74] in January) and Ramchandani ([Ram74] in February) about the combination of time and Petri nets appeared almost simultaneously. Merlin uses Petri nets to study the formal analysis and synthesis of recoverability of communication protocols. He extends Petri nets to overcome certain practical restrictions by introducing time in the following way: An interval $[a_t, b_t]$ is assigned to every transition t. The firing rule is modified in relation to time: An enabled transition t cannot fire immediately after being enabled. At least a_t time units must pass before t may fire and it must fire no later than b_t time units after being enabled, except if it becomes disabled in the meantime. The times a_t and b_t are considered relative to the latest enabling of t. Firing itself takes no time. Merlin called these time-dependent nets *Time Petri nets*.

Timed Petri nets were introduced by Ramchandani to model speeds of operations or of parts of processes. He assigns to every transition t a duration d_t. In a Timed Petri net if a transition t is enabled, it has to fire immediately. Thereby a maximum number of (just) enabled transitions is always fired, i.e., they are fired according to the *maximal-step rule*. The firing of a transition t lasts d_t time units and cannot be interrupted.

Petri nets with indication of time in the places were introduced by Sifakis [Sif77]. He assigns a minimum retention time d_p to every place p. If a token reaches place p at time τ, it must remain there for at least d_t time units, before a transition is allowed to fire using this token. The transitions in these nets fire according to the maximal-step rule.

Finally, *Petri nets with time-dependent arcs* were introduced at the beginning of the 1980s in [Wal82]. A period of time τ_k is associated with each arc k. A transition t can fire in such a net if the number of tokens on every input place p of t is at least as great as the multiplicity of the arc (t, q). These tokens must remain on p for $\tau_{(p,t)}$ time units before they are used for firing. Then, $\tau_{(t,q)}$ time units after the firing of transition t to each post-place q of t, the number of tokens corresponding to the multiplicity of the arc (t, q) is added.

In the course of the last two decades all kinds of variations and combinations of the time-dependent Petri nets specified above have arisen. Some of them can be translated into each other. The introduction of each new class is nevertheless justified if it is a natural means of specifying practical systems. A study about the expressive power of Time Petri nets and their extensions as well as a comparison to timed automata is done in [BCH+13].

In addition to Petri nets with deterministic time extensions, time-dependent Petri nets in which firing occurs according to a random distribution over time have been examined, see [MBC$^+$96], [BK02] etc. These nets are called stochastic Petri nets. An essential difference between stochastic and non-stochastic Petri nets is the relationship between the state spaces of the time-dependent Petri net and that of the underlying timeless Petri net: A stochastic Petri net has the same state space as the underlying Petri net. The state space of a deterministic time-dependent Petri net on the other hand in general comprises only part of the state space of the underlying timeless Petri net. This alone leads to different analysis algorithms. For the analysis of deterministic Petri nets, for example, it is essential to try to determine the state space.

In this book we will look at the first three classes of Petri nets with time extensions mentioned above, Time Petri nets, Timed Petri nets and Petri nets with retention time in the places. We will introduce new methods of analysis for these types of nets. Petri nets with time-dependent arcs can be reduced to these three classes. An important aspect of the analysis is the description of the state space. We can consider a complete, reduced, or parametric state space. Fundamental properties such as boundedness and liveness are defined anew, under consideration of the time specification, and methods of analysis are introduced. The behavior of time-dependent and timeless nets is compared and algorithms for qualitative and quantitative analysis are presented. Applications of these nets are mentioned together with their use in specification and analysis.

This book is structured as follows: In Chapter 2 classic Petri nets without time are introduced. In the subsequent chapters Time Petri nets and Timed Petri nets as well as Petri nets with time-dependent places (Petri nets with time windows) are studied.

Notation

In this book the set of all natural numbers is denoted by $\mathbb{N}$ whereas $\mathbb{N}^+$ stands for the set of all natural numbers excluding 0. $\mathbb{Q}_0^+$ stands for the non-negative rational numbers and $\mathbb{R}_0^+$ for the set of non-negative real numbers.

$\mathbb{N}^n$ denotes the n-ary Cartesian product over the set of natural numbers $\mathbb{N}$.

Let u and v be vectors of dimension n. Then u is less than or equal to v ($u \leq v$) if every component of u is less than or equal to the corresponding component of v. The sum $u + v$ of the two vectors u and v is also a vector of dimension n, whose components are the sums of the corresponding components of u and v. The difference $u - v$ is defined analogously.

Let A be a finite set. Then A^* is the set of all finite words (sequences) over A. ε is the empty word. $\ell(w)$ denotes the length of the word w. A^+ stands for the set $A^* \setminus \{\varepsilon\}$. The number of elements of A is denoted by $|A|$.

For an arbitrary function $f : A \longrightarrow B$ the set A is called the domain and B the codomain of f.

An n-ary function f is called arithmetical or number-theoretical if its domain is the set $\mathbb{N}^n$ and its codomain is the set $\mathbb{N}$.

Furthermore W^D is defined as the set of all functions with domain D and codomain W.

Finally, the real number r rounded down is written as $\lfloor r \rfloor$ and r rounded up as $\lceil r \rceil$.

Chapter 2

The Classic Petri Net

In this chapter the basic principles of classic Petri net theory are introduced and basic properties are explained. With the latter, the emphasis will not be on a complete listing but on the systematization of qualities as static or dynamic and qualitative or quantitative.

2.1 Definitions

Definition 2.1 (unmarked Petri net) *An unmarked Petri net is a 4 tuple* (P, T, F, V) *such that*

1. *P and T are finite sets with $P \cap T = \emptyset$ and $P \cup T \neq \emptyset$.*

2. *F is a relation of arity 2 with $F \subseteq (P \times T) \cup (T \times P)$.*

3. *$V : F \longrightarrow \mathbb{N}^+$.*

The elements of P are called *places* and the elements of T are called *transitions*. The elements of F are called *arcs* and F is called the *flow relation* of $\mathcal{N}$. The function V is the *multiplicity (weight)* of the arcs.

This definition covers the static aspects of a Petri net. An unmarked Petri net is therefore a 2-colored, weighted, directed, finite graph. The vertices of one color represent the places and the vertices of the other color the transitions.

In Petri net theory, the places are graphically represented by circles and the transitions by rectangles or bars.

Definition 2.2 (Petri net) *A marked Petri net is a 5-tuple $\mathcal{N} = (P, T, F,$ $V, m_0)$ (short: Petri net) such that*

1. *(P, T, F, V) is an unmarked Petri net.*

2. *$m_0 : P \longrightarrow \mathbb{N}$ is the initial marking[1].*

Thus, a Petri net has an initial marking which assigns a natural number to each place. This marking is graphically represented by the corresponding number of tokens (points) on the places, so $m_0(p)$-tokens are drawn in the circle representing the place p. Any distribution of tokens on the places is a marking:

Definition 2.3 (marking) *Let P be the set of places of a Petri net $\mathcal{N}$. A marking in $\mathcal{N}$ is a total function $m : P \longrightarrow \mathbb{N}$.*

Example 2.4

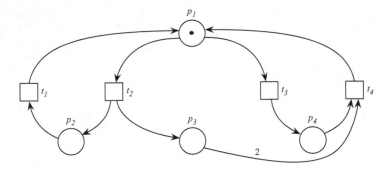

Figure 2.1: $\mathcal{N}_1$ is a Petri net with the initial marking $m_0 = (1, 0, 0, 0)$

---

[1]The function m_0 is a total one.

The initial marking of a Petri net can generally change into a successor marking according to certain rules and this can itself transform in turn into successor markings. The rules describing the possible changes from one marking to the next one are called *firing rules*, the occurring change itself is called a *firing*. Throughout such firings, the distribution of tokens over the places of a Petri net can change and thereby the whole view of the net changes. In other words: The Petri net also has a *dynamic* aspect which is defined by the firing rules.

Before we define the firing rules for classic Petri nets, some basic static notions need to be explained: Let $\mathcal{N}$ be an arbitrary Petri net with a set of places P, a set of transitions T and a flow relation F. All places which are connected to a transition by an arc form the set of pre-places and post-places of a specific transition. A pre-place of a certain transition is a place which is directly connected with the considered transition through an arc directed from the place to the transition. If the arc points in the opposite direction, that place is a post-place of the transition. The *set of pre-places* of the transition t is denoted by ${}^\bullet t := \{p \mid p \in P \wedge (p, t) \in F\}$. The set $t^\bullet := \{p \mid p \in P \wedge (t, p) \in F\}$ is the *set of post-places* of t. Analogously, for every place p the set ${}^\bullet p := \{t \mid t \in T \wedge (t, p) \in F\}$ denotes the *set of pre-transitions* of p and the set $p^\bullet := \{t \mid t \in T \wedge (p, t) \in F\}$ denotes the *set of post-transitions* of p.

The dynamic aspect of a Petri net is defined by the firing rules. The firing rules reflect causal relations within a permanently changing system: The events of the real system are modeled by transitions of the Petri net. The causes or preconditions of an event are represented by the pre-places of the transition modeling the event. The post-places of the transition describe the post-conditions of the event, which of course in turn can be preconditions of other events. Whenever a pre-place is marked, the respective condition is considered to be fulfilled. In the real system an event can take place provided that all preconditions of the event are fulfilled. In the Petri net the occurrence of the event is represented by firing of the respective transition. After an event has taken place, its preconditions (in general) are not fulfilled any more. The corresponding pre-places are therefore no longer marked. Instead the post-conditions of the event are fulfilled and in the Petri net the post-places of the transition are marked. This atomic process in the real system provides the basic idea for the firing rule in the Petri net. In classic Petri nets this basic idea is carried out with the multiplicity $V \equiv 1$ (these are the so-called *ordinary Petri nets*) and each place holds at most one token, i.e., for each reachable marking m (cf. Def. 2.10) it holds that $m : P \longrightarrow \{0,1\}$ (these are the so called *1-safe Petri nets*). Ordinary, 1-safe Petri nets are also called *condition/event nets*.

If not all multiplicities of arcs of a Petri net are 1, the firing rule is extended consistently. The preconditions of an event are fulfilled if, for each place, the number of tokens it holds is no smaller than the multiplicity of the arc from this place to the respective transition. After an event has taken place, every post-place of the transition obtains the number of tokens equivalent to the multiplicity of the arc from the transition to the respective post-place. A transition t can therefore fire if the Petri net is in a marking m, which assigns at least as many tokens as t needs in each pre-place of t. All preconditions can be regarded as fulfilled. This minimum number of necessary tokens on the places is defined by the marking t^-:

$$t^-(p) := \begin{cases} V(p,t) & , \text{ if } (p,t) \in F \\ 0 & , \text{ if } (p,t) \notin F \end{cases}.$$

Analogously the marking t^+ describes the number of tokens which are added to each place upon firing of t:

$$t^+(p) = \begin{cases} V(t,p) & \text{ if } (t,p) \in F \\ 0 & \text{ if } (t,p) \notin F \end{cases}.$$

The difference in tokens on the places after firing of transitions t is represented by the marking Δt :

$$\Delta t := t^+ - t^-.$$

Furthermore, let us consider the Petri net $\mathcal{N} = (P, T, F, V, m_0)$ with $T = \{t_1, \cdots, t_n\}$ and $P = \{p_1, \cdots, p_m\}$. The matrix $C_{\mathcal{N}} = (c_{ij})$ with m rows and n columns and

$$c_{ij} := \Delta t_j(p_i)$$

is called the *incidence matrix* of $\mathcal{N}$.

Example 2.5 *We consider again the Petri net $\mathcal{N}_1$ introduced in Example 2.4. Its incidence matrix $C_{\mathcal{N}_1}$ is:*

$$C_{\mathcal{N}_1} = \begin{pmatrix} 1 & -1 & -1 & 1 \\ -1 & 1 & 0 & 0 \\ 0 & 1 & 0 & -2 \\ 0 & 0 & 1 & -1 \end{pmatrix} \begin{matrix} P_1 \\ P_2 \\ P_3 \\ P_4 \end{matrix} .$$

$$\Delta t_1 \ \Delta t_2 \ \Delta t_3 \ \Delta t_4$$

We can now formally introduce the notions *enabled* and *firing*:

Definition 2.6 (enabled) *Let $\mathcal{N} = (P, T, F, V, m_0)$ be a Petri net and let m be a marking in $\mathcal{N}$. A transition $t \in T$ is enabled in m if it holds that: $t^- \leq m$.*

Definition 2.7 (firing) *Let $\mathcal{N} = (P, T, F, V, m_0)$ be a Petri net and let m be a marking in $\mathcal{N}$. A transition $t \in T$ can fire in m (notation: $m \xrightarrow{t}$), if t is enabled in m. After the firing of t the Petri net is in the marking m' (notation: $m \xrightarrow{t} m'$) with*

$$m' := m + \Delta t.$$

This firing rule defines the firing of *a single* transition, i.e., it is a *mono*-firing rule. It is also possible to define a rule firing a set (step) of transitions. This firing rule, also called a *step firing rule*, is usually used in time-dependent Petri nets, see Chapter 4. We will always use the mono-firing rule unless stated otherwise.

We denote the change of a Petri net from marking m into marking m' by firing of transition t by $m \xrightarrow{t} m'$. The relation $\longrightarrow$ which is defined by the firing rule is called the *firing relation*.

A further basic term in Petri Net theory is the notion of a *reachable marking*. In order to introduce it, we first define *firing sequences*:

Definition 2.8 (firing sequence) *Let $\mathcal{N} = (P, T, F, V, m_0)$ be a Petri net, let m be a marking in $\mathcal{N}$ and let $\sigma = t_1 \cdots t_n$ be a sequence of transitions. σ fires from m to m' in $\mathcal{N}$ (short: $m \xrightarrow{\sigma} m'$) if it holds that:*

Basic $\sigma = \varepsilon$

$\quad m' := m$

Step $\sigma = t_1 \cdots t_n t_{n+1}$

There is a marking m'' in $\mathcal{N}$, with

$$m \xrightarrow{t_1 \cdots t_n} m'' \quad and \quad m'' \xrightarrow{t_{n+1}} m'.$$

σ is called a firing sequence from m in $\mathcal{N}$ if there is a marking m' such that σ fires from m to m'.

A firing sequence σ from m_0 in $\mathcal{N}$ is usually simply called a firing sequence.

Example 2.9 *Let us consider example 2.4 again. The transitions sequence* $\sigma_1 = t_2\,t_1 t_2 t_1 t_3 t_4$ *is a firing sequence from* m_0 *in the Petri net* $\mathcal{N}_1$. *In contrast, the transition sequence* $\sigma_2 = t_3 t_4 t_3 t_4$ *is not a firing sequence from* m_0 *in the same net. However,* σ_2 *is a firing sequence from* $m = (1, 0, 4, 0)$ *in* $\mathcal{N}_1$.

We write $m \xrightarrow{*} m'$ if there exists a firing sequence σ such that $m \xrightarrow{\sigma} m'$. This means that the relation $\xrightarrow{*}$ is the reflexive-transitive closure of the firing relation $\longrightarrow$.

Furthermore, for $\sigma = t_1 \cdots t_n$ the following is true:

$$m' = m + \sum_{i=1}^{n} \Delta t_i. \tag{1}$$

This equation can equivalently be rewritten in the form:

$$m' = m + \sum_{t \in \sigma} \pi_t \cdot \Delta t, \tag{2}$$

where π_t is the number of appearances of the transition t in the sequence σ. Finally, we obtain the equality

$$m' = m + C_{\mathcal{N}} \cdot \pi_\sigma \tag{3}$$

where $C_{\mathcal{N}}$ is the incidence-matrix of $\mathcal{N}$ and $\pi_\sigma \in \mathbb{N}^T$ the vector with

$$\pi_\sigma(t) = \begin{cases} \pi_t & \text{if } t \in \sigma \\ 0 & \text{otherwise.} \end{cases}$$

The $|T|$-dimensional vector π_σ is called *the Parikh vector* of σ. The equality (3) is called the *state equation of σ in m (in $\mathcal{N}$).*

Definition 2.10 (reachable marking) *A marking m is called reachable from the marking m^* in a Petri net $\mathcal{N}$, if there is a firing sequence σ from m^* to m in $\mathcal{N}$. If $m^* = m_0$ we call m reachable in $\mathcal{N}$.*

Finally, $R_{\mathcal{N}}(m) := \{m' \mid m \xrightarrow{*} m'\}$ is the notation for the set of all markings m' which are reachable from the marking m in $\mathcal{N}$.

2.2   State Space

For any Petri Net $\mathcal{N}$ the set $R_{\mathcal{N}}(m_0)$ contains all markings reachable in the net. This set is of particular interest because it gives us information about all the events that can occur in a system modeled by the considered net. The set tells us which of the markings of the net are reachable and thereby also which pre-conditions of events may potentially be fulfilled.

Definition 2.11 (state space) *Let $\mathcal{N} = (P, T, F, V, m_0)$ be a Petri net. The set $R_{\mathcal{N}} := R_{\mathcal{N}}(m_0)$ is called the state space of $\mathcal{N}$.*

The state space can be finite or infinite. The fact that the set $R_{\mathcal{N}}$ is decidable (cf. inter alia [PW03]) is crucial for the analysis of Petri nets as the state space holds information about the reachability/non-reachability of markings in $\mathcal{N}$ and thus about the occurrence/non-occurrence of events.

Definition 2.12 (boundedness) *A Petri net $\mathcal{N}$ is said to be bounded if the set $R_{\mathcal{N}}$ of all its reachable markings is finite.*

Boundedness can also be introduced using the notion of *bounded place*. A place in a Petri net is bounded if there is a natural number such that the number of tokens on this place never exceeds this number. A net is then bounded if all its places are bounded. These definitions of boundedness are equivalent.

We now consider again the reflexive-transitive closure of the firing relation. This relation generally fulfills none of the common properties of relations like symmetry, asymmetry, anti-symmetry, connexity etc. The graph of the relation is called *the reachability graph* of the Petri net. It is formally defined as follows:

Definition 2.13 (reachability graph) *Let $\mathcal{N} = (P, T, F, V, m_0)$ be a Petri net. The reachability graph of $\mathcal{N}$ is the graph $\mathcal{RG}_\mathcal{N}$ with*

1. *the set $R_\mathcal{N}$ as set of vertices and*

2. *$(m, m') \in R_\mathcal{N} \times R_\mathcal{N}$ is an edge in $\mathcal{RG}_\mathcal{N}$ if there is a transition $t \in T$ such that $m \xrightarrow{\ t\ } m'$.*

Such a reachability graph of a Petri net is a partial deterministic automaton which can be finite or infinite (also compare Fig.2.2). The bounded Petri nets are furthermore exactly those Petri nets whose reachability graphs are finite. In the case of boundedness the Petri net is therefore well analyzable with the help of its reachability graph. Determining whether a certain property holds in an unbounded Petri net however is a lot more challenging.

Example 2.14 *Let us again consider the Petri net $\mathcal{N}_1$ given in Example 2.4. Its reachability graph $\mathcal{RG}_{\mathcal{N}_1}$ is infinite. A part of the reachability graph is presented in Fig.2.2.*

When a Petri net is unbounded its reachability graph is infinite. In this case we can consider the so-called *coverability graph* of the net, which is always finite, to prove the existence or absence of properties. The trade-off for the finiteness of the coverability graph is loss of information.

The vertices of the coverability graph are the so-called *generalized markings* of the Petri net. A generalized marking in a Petri net is a (total) function which assigns to each place a natural number or the value ω. A place has the value ω in a generalized marking when the place is unbounded. We extend the rules of addition, subtraction, and multiplication and the relation $\leq$ for the set $\mathbb{N} \cup \{\omega\}$ as follows:

For each $n \in \mathbb{N}$:

$$\omega \pm n = \pm n + \omega = \omega, \quad n \cdot \omega = \omega \cdot n = \begin{cases} \omega & \text{if } n > 0 \\ 0 & \text{if } n = 0 \end{cases}, \quad \omega > n.$$

We say that a (generalized) marking m *covers* another (generalized) marking m' in a Petri net $\mathcal{N}$ (short: $m' \prec m$), if $m' \leq m$ and there is at least one place p in $\mathcal{N}$ with $m(p) \neq m'(p)$. Therefore, $m' \leq m$ implies that either $m' \prec m$ or $m' = m$.

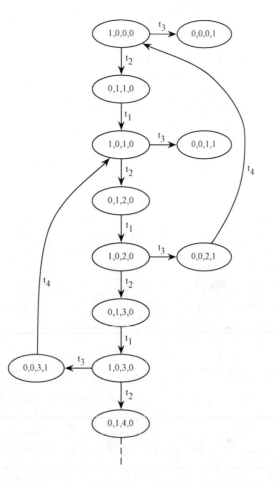

Figure 2.2: Part of the reachability graph $\mathcal{RG}_{\mathcal{N}_1}$

Definition 2.15 (coverability graph) *Let* $\mathcal{N} = (P, T, F, V, m_0)$ *be a Petri net. The edge-labeled digraph* $\mathcal{CG}_{\mathcal{N}} := (W, E, T)$ *is said to be a coverability* graph of $\mathcal{N}$ if the set of vertices W, the set of edges E and the set of labels T are defined using the following algorithm [2]:

begin $R := \{m_0\};$ $ancestor\_marking(m_0) := *;$ $W := \emptyset;$ $E := \emptyset;$

    while $R \neq \emptyset$ **do**

        choose m from $R;$ $R := R - \{m\};$ $W := W \cup \{m\};$

        $enabled\_set := \{t \mid t^- \leq m\};$

        for $t \in enabled\_set$ **do**

            $m' := m + \Delta t;$

            $m^* := m;$

            while $(m^* \neq *)$ and $(m^* \not\leq m')$ **do**

                $m^* := ancestor\_marking(m^*);$

            end;

            if $m^* \neq *$ **then**

                $m' := m' + (m' - m^*) \cdot \omega;$

            end;

            $E := E \cup \{(m, t, m')\};$

            if $m' \notin W \cup R$ **then**

                $R := R \cup \{m'\};$ $ancestor\_marking(m') := m;$

            end;

        end;

    end;

end.

The coverability graph is not unique for a given Petri net. The notion "coverability graph" was first defined by Karp and Miller (see [KM69]). Later, in [Fin93], Finkel introduced an algorithm which computes a minimal coverability graph of a Petri net. This graph is unique and has the minimum number of vertices.

---

[2]Adapted from [Sta90]

Example 2.16 *Using Definition 2.15 we obtain for the Petri net $\mathcal{N}_1$ considered in Example 2.4 the coverability graph $\mathcal{CG}_{\mathcal{N}_1}$, represented in Fig. 2.3:*

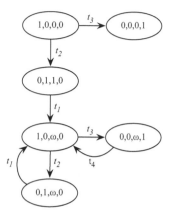

Figure 2.3: A coverability graph $\mathcal{CG}_{\mathcal{N}_1}$ of the Petri net $\mathcal{N}_1$

2.3 PN-Computability

The expressive power of Petri nets is less than that of Turing machines (TM). This means that there are algorithms which cannot be described by any classic Petri net: Not every Turing-computable function can also be computed by a Petri net.

A number-theoretical n-ary function $f : \mathbb{N}^n \longrightarrow \mathbb{N}$ is called *Turing-computable* if there is a Turing machine M_f which given an n-tuple $(x_1, \cdots, x_n)$ as input stops if and only if that n-tuple belongs to the domain of the function and in this case also returns the value $f(x)$. For more on Turing-computability compare [HMU02].

We now need to clarify what a *PN-computable function* is. Regardless of how we define that notion, we first have to ensure that there is a unique presentation for every natural number in a Petri net. The infinity of the set of all natural numbers prevents it from being modeled by places, transitions or arcs. But the set of reachable markings in a Petri net can in general be infinite. Moreover, transitions might fire infinitely often and places can change their number of tokens infinitely often, even if the places themselves are bounded.

In order for a place to change its number of tokens infinitely often, at least one transition in the net needs to fire infinitely often. The reverse is also true and therefore the two properties are equivalent.

It can furthermore be shown that the number of firings in an arbitrary Petri net is equal to the number of reachable markings in another Petri net derived from the first one as follows:

Let $\mathcal{N} = (P, T, F, V, m_0)$ be an arbitrary Petri net. Using $\mathcal{N}$ we construct the Petri net $\mathcal{N}' = (P', T', F', V', m_0')$ where:

$$P' := P \cup \{p^*\} \text{ with } p^* \notin P, \quad T' := T, \quad F' := F \cup \{(t, p^*) \mid t \in T\} \qquad \text{and}$$

$$V'(u) := \begin{cases} V(u) & \text{if } u \neq (t, p^*) \\ 1 & \text{if } u = (t, p^*) \end{cases}, \quad m_o'(p) := \begin{cases} m_0(p) & \text{if } p \neq p^* \\ 0 & \text{if } p = p^* \end{cases}$$

for all $p \in P'$ and for all $t \in T'$.

The net $\mathcal{N}'$ is a copy of the net $\mathcal{N}$ with one additional place p^* such that every time a transition in the copy of $\mathcal{N}$ fires, the number of tokens in p^* increases by 1. Thus the place p^* in $\mathcal{N}'$ is unbounded if and only if there is at least one transition t in the net $\mathcal{N}$ which fires infinitely often.

It is thereby clearly possible to represent natural numbers – and therefore also n-tuples of natural numbers – by markings in a Petri net.

Next we define the notion of *PN-computability*:

Definition 2.17 (PN-computable) *An n-ary function $f : \mathbb{N}^n \longrightarrow \mathbb{N}$ is called Petri-net-computable (PN-computable) if there is an initial Petri net[3] $\mathcal{N}_f = (P_f, T_f, F_f, V_f, m_0^f)$ such that for each n-tuple $x = (x_1, \cdots, x_n) \in \mathbb{N}^n$*

---

[3]An initial Petri net is an arbitrary Petri net which is fixed for a function f. The extension of the initial marking m_0^f of $\mathcal{N}_f$ to initial markings $m_0^{f,x}$ modeling the arguments $x = (x_1, \cdots, x_n)$ generates the Petri nets $\mathcal{N}_f^x$. The marking m_0^f is better understood after reading the similar definition for Time Petri nets, Definition 3.15, as well as Examples 3.17 and 3.18.

and for the Petri net $\mathcal{N}_f^x = (P_f, T_f, F_f, V_f, m_0^{f,x})$, where the marking $m_0^{f,x}$ models the n-tuple $(x_1, \cdots, x_n)$, it holds that:

Case 1 *If the tuple $(x_1, \cdots, x_n)$ belongs to the domain of f then the Petri net $\mathcal{N}_f^x$ stops (cannot fire anymore) and the last-reached marking $m^{f(x_1,\cdots,x_n)}$ uniquely represents the number $f(x_1, \cdots, x_n)$.*

Case 2 *If the tuple $(x_1, \cdots, x_n)$ does not belong to the domain of f then the Petri net $\mathcal{N}_f^x$ never stops, i.e., for each marking m with $m_0^{f,x} \xrightarrow{*} m$ there exists at least one transition t which is enabled in m.*

Evidently, every PN-computable function is also Turing-computable.

We *assume* now that every Turing-computable function is also PN-computable. Let f be an arbitrary, n-ary, Turing-computable function. According to the assumption f is PN-computable, too. Let M_f be a Turing machine which computes f and $\mathcal{N}_f$ a Petri net which computes f. Furthermore, let $(x_1, \cdots, x_n)$ be an n-tuple of natural numbers. Then it holds that:

$$M_f \text{ started on } (x_1, \cdots, x_n) \text{ stops}$$
$$\text{if and only if}$$
$$\mathcal{N}_f \text{ started in } (x_1, \cdots, x_n) \text{ stops.} \tag{4}$$

As detailed before, we can define the Petri net $\mathcal{N}'$ for every Petri net $\mathcal{N}$. The construction of the respective Petri net $\mathcal{N}_f'$ for $\mathcal{N}_f$ ensures that for any $\mathcal{N}_f$ the following holds:

$$\mathcal{N}_f \text{ started in } (x_1, \cdots, x_n) \text{ stops}$$
$$\text{if and only if}$$
$$\text{the place } p^* \text{ in } \mathcal{N}_f' \text{ is bounded.} \tag{5}$$

We can determine whether the place p^* is bounded in $\mathcal{N}_f'$ by means of the coverability graph of $\mathcal{N}_f'$. Taking into account (4) and (5) it is consequently decidable for an arbitrary Turing-computable function f whether the Turing machine M_f started with an arbitrary n-tuple $(x_1, \cdots, x_n)$ stops or not. This contradicts, however, the undecidability of the halting problem for Turing machines (cf. among others [HMU02]). Therefore the assumption that every Turing-computable function is also PN-computable is disproved.

Thus we have proved the property:

Theorem 2.18 *The class of all marked Petri nets (as defined in Definition 2.2 and using the mono-firing rule) is not Turing-complete.*

2.4   Basic Properties

The potential of Petri nets for modeling results from their graphical representability and makes them applicable in a wide variety of areas. Many kinds of systems can be represented easily but nevertheless clearly and formally using Petri nets. Furthermore, the nature of Petri nets allows modeling of a system at each stage of abstraction and refinement of the model where necessary. Therefore modeling with Petri nets is a hierarchical procedure as well as a modular one. Without analyzing methods for Petri nets however their use as a modeling instrument is very limited. We want on the one hand to ascertain the presence of certain properties in a Petri net in order to evaluate its "quality" as a model of a real system, on the other hand an analysis of the model might detect properties of the real system previously overlooked.

Some properties of Petri nets can be determined very easily, these in general being properties which result from the structure of the Petri net itself. We call such properties *static*. In contrast to this, we describe properties which result from firing and which characterize permanent changes in the net as *dynamic* properties. These dynamic properties are usually more difficult to identify and sometimes impossible to analyze.

Among the static properties there are such features as the existence of static conflicts, deadlocks, traps, etc., or whether a net is a marked graph, a state machine, a free-choice net, an extended-free-choice net, an asymmetric-choice net, a homogeneous one, etc. All of these properties are decidable independently of the state space of the considered Petri nets. Some of them are relevant for our studies and will be defined below, although they are not the main subject of our analysis.

The notions *free-choice net, extended-free-choice net, asymmetric-choice net* and *marked graph* were originally defined for ordinary Petri nets and are still mainly used in that context. For arbitrary multiplicity these notions were first introduced in [Sta90]. In this book, we use more recent and more general definitions. Therefore the classes defined here, like free-choice nets, extended-free-choice nets, asymmetric-choice nets and marked graphs for homogeneous Petri nets, are supersets of the corresponding classes of ordinary nets.

Definition 2.19 *Let $\mathcal{N} = (P, T, F, V, m_0)$ be a Petri net.*

1. *Two transitions t and t' from T are in a* **static conflict***, if they have at least one common pre-place, i.e., ${}^\bullet t \cap {}^\bullet t' \neq \emptyset$.*

2. *Two transitions t and t' from T are in a* **dynamic conflict** *in the marking m if they are in a static conflict and by firing one of the transitions in marking m the other one may become disabled, i.e., ${}^\bullet t \cap {}^\bullet t' \neq \emptyset$ and $t^- \leq m$ and $t'^- \leq m$ but $t^- + t'^- \not\leq m$.*

3. *$\mathcal{N}$ is a* **marked graph** *if $\mathcal{N}$ is an ordinary Petri net and if each place p has exactly one pre-transition and one post-transition, i.e., $V(f) = 1$ for each $f \in F$ and $|{}^\bullet p| = |p^\bullet| = 1$.*

4. *$\mathcal{N}$ is a* **free-choice net** *(FC net) if each shared place is the only pre-place of its post-transitions, i.e., if $t, t' \in p^\bullet$ then ${}^\bullet t = \{p\} = {}^\bullet t'$.*

5. *$\mathcal{N}$ is an* **extended-free-choice net** *(EFC net) if the post-transitions of each shared place have the same pre-places, i.e., if $t, t' \in p^\bullet$ then ${}^\bullet t = {}^\bullet t'$.*

6. *$\mathcal{N}$ is an* **asymmetric-choice net** *(AC net) if it holds that if two places have at least one common post-transition then the set of all post-transitions of one of the places is a subset of the set of all post-transitions of the other one. Formally: For each two places p and p' it holds: If $p^\bullet \cap p'^\bullet \neq \emptyset$ then $p^\bullet \subseteq p'^\bullet$ or $p'^\bullet \subseteq p^\bullet$.*

7. *$\mathcal{N}$ is* **homogeneous** *if all output arcs of a place have the same multiplicity, i.e., for each place $p \in P$ it is true that if $t, t' \in p^\bullet$, then $V(p, t) = V(p, t')$.*

Example 2.20 *Let us consider the Petri net $\mathcal{N}_1$ from Example 2.4 . The transitions t_2 and t_3 are in a static conflict. $\mathcal{N}_1$ is a FC net and therefore also an EFC net and an ES net. It is homogeneous.*

We refrain from defining further static properties and refer to the common literature on classic Petri nets, such as [Mur89] or [Sta90].

Liveness, boundedness, reachability, the existence of invariants, etc. are the basic properties of a Petri net and are dynamic properties. They provide information about the behavior of the examined Petri net.

The notion of *liveness* was established by Petri, Genrich and Lautenbach at the GMD, Bonn at the beginning of the 1960s and developed further in cooperation with Commoner, Even, Holt, Pnueli and Hack at MIT. It was initially examined in special classes of Petri nets such as marked graphs (cf. [Gen68], [CHEP71]) and AC-nets (cf. [Hac72], [Com73]). The *four levels* of liveness introduced by Lautenbach in his PhD thesis [Lau73] in 1973 constituted an important advance and form the basis of liveness studies up to the present day.

Definition 2.21 *Let $\mathcal{N} = (P, T, F, V, m_0)$ be a Petri net, m be an arbitrary marking in $\mathcal{N}$ and t a transition in T.*

1. *t is called **live in** m in $\mathcal{N}$ if for each marking $m' \in R_{\mathcal{N}}(m)$ there exists a marking $m'' \in R_{\mathcal{N}}(m')$ such that $m'' \xrightarrow{t}$.*

2. *t is called **dead in** m in $\mathcal{N}$ if for each marking $m' \in R_{\mathcal{N}}(m)$ it holds that $t^- \not\leq m'$.*

3. *m is called **live in** $\mathcal{N}$ if each transition $t \in T$ is live in m.*

4. *m is called **dead in** $\mathcal{N}$ if each transition $t \in T$ is dead in m.*

5. *t is called **live / dead in** $\mathcal{N}$ if t is live / dead in m_0.*

6. *$\mathcal{N}$ is called **live / dead** if m_0 is live / dead in $\mathcal{N}$.*

7. *$\mathcal{N}$ is called **blocking-free**[4] if at least one transition is enabled in each reachable marking $m \in R_{\mathcal{N}}(m_0)$.*

As is obvious from the definition, a transition t is live in the marking m if it is not dead in any successor marking of m. If a marking is dead, no transition can fire from that marking. The marking therefore constitutes a leaf in the

---

[4]Some authors call this property *deadlock-free*.

reachability graph of the net. Furthermore we observe that a Petri net which is not dead does not have to be blocking-free (deadlock-free). The reason for this being that a Petri net which fires only finitely often and then stops is not dead. A blocking-free Petri net obviously is not dead. Every live Petri net is blocking-free, the reverse however does not hold. A Petri net which is not dead need not be live and a Petri net which is not live is not necessarily dead.

We can summarize as follows:

$$\mathcal{N}- \text{dead} \quad \xrightarrow{\quad}_{\;\not\leftarrow\;} \quad \mathcal{N}- \text{not blocking-free} \quad \xrightarrow{\quad}_{\;\not\leftarrow\;} \quad \mathcal{N}- \text{not live}$$

and

$$\mathcal{N}- \text{live} \quad \xrightarrow{\quad}_{\;\not\leftarrow\;} \quad \mathcal{N}- \text{blocking-free} \quad \xrightarrow{\quad}_{\;\not\leftarrow\;} \quad \mathcal{N}- \text{not dead.}$$

Whether a Petri net is live, dead or blocking-free can be decided by means of its reachability graph. The coverability graph of a Petri net however is of no use in deciding liveness and blocking-freedom but does still contain the information of whether the net is dead (cf. [PW03]). There are a number of algorithms which decide these properties more effectively on restricted classes of nets.

Finally we remark that the notion of t *being live in a marking* m introduced in Definition 2.21 is equivalent to the *4-liveness* defined by Lautenbach in [Lau73]. The notion of t *being dead in a marking* m is equivalent to the *0-liveness* also defined in [Lau73]. In other words, t is dead in m according to Definition 2.21 if and only if t is not 1-live in m. This analogously applies to the liveness of a marking and to the liveness of a Petri net. Finally, at least one transition in a Petri net is *3-live* as defined by Lautenbach if the Petri net is blocking-free. The converse does not hold.

We already introduced the notions of *boundedness* and *reachable markings* with Definitions 2.12 and 2.10.

Example 2.22 *Let us consider the Petri net* $\mathcal{N}_1$ *from Example 2.4 again.*

- *The place* P_3 *is unbounded.*

- $\mathcal{N}_1$ *is unbounded.*

- $\mathcal{N}_1$ *is not live.*

- $\mathcal{N}_1$ *is not dead.*

- $\mathcal{N}_1$ *is not blocking-free.*

With the last two definitions in this chapter we will now introduce two (dual) notions of invariants in Petri nets:

Definition 2.23 (T-invariant) *Let $C_\mathcal{N}$ be the incidence matrix of the Petri net $\mathcal{N}$.*

1. *Each non-trivial solution $x \in \mathbb{N}^{|T|}$ of the homogeneous equality*

$$C_\mathcal{N} \cdot x = 0$$

   is called a transitions-invariant (short: T-invariant) of $\mathcal{N}$.

2. *A Parikh vector of a firing sequence which is also a T-invariant is called a feasible T-invariant.*

It is not difficult to see that if a Parikh vector π_σ is a T-invariant the marking is not changed by the firing of the sequence σ. This follows from the state equality for σ:

$$m \xrightarrow{\sigma} m'$$
$$\text{iff}$$
$$m' = m + C_\mathcal{N} \cdot \pi_\sigma = m$$
$$\text{i.e.,}$$
$$m' = m.$$

The path representing the sequence σ in the reachability graph is obviously a cycle.

Example 2.24 *Let us consider the Petri net $\mathcal{N}_1$ shown in Example 2.4. The integer solutions of the homogeneous equality system*

$$C_{\mathcal{N}_1} \cdot \begin{pmatrix} x_1 \\ x_2 \\ x_3 \\ x_4 \end{pmatrix} = \begin{pmatrix} 0 \\ 0 \\ 0 \\ 0 \end{pmatrix} \tag{6}$$

are $\quad x_1 = x_2 = 2 \cdot k, \quad x_3 = x_4 = k \quad$ for $\ k \in \mathbb{N}$.

Furthermore, we note that the transition sequence $\sigma = t_2 t_1 t_2 t_1 t_3 t_4$ is a firing sequence starting in m_0 and its Parikh vector π_σ is a solution of the equality (6).

$\overline{We}$ can see in Example 2.14 that the path σ is a cycle in the reachability graph $\mathcal{RG}_{\mathcal{N}_1}$.

Definition 2.25 (P-invariant) Let $C_{\mathcal{N}}$ be the incidence matrix of the Petri net $\mathcal{N}$. Each non-trivial solution $y \in \mathbb{N}^{|P|}$ of the homogeneous equality

$$y^T \cdot C_{\mathcal{N}} = 0$$

is called a place-invariant (short: P-invariant) of $\mathcal{N}$.

We can view a P-invariant as an equation of weights for the places that always holds:

Let $y = (y_1, \cdots, y_{|P|})$ be a P-invariant in a Petri net $\mathcal{N}$. It is true for each reachable marking m in $\mathcal{N}$ that :

$$m = m_0 + C_{\mathcal{N}} \cdot \pi_\sigma$$

and therefore

$$y^T \cdot m = y^T \cdot m_0 + y^T \cdot C_{\mathcal{N}} \cdot \pi_\sigma$$

and thus subsequently

$$\sum_{i=1}^{|P|} y_i \cdot m(p_i) = const. = \sum_{i=1}^{|P|} y_i \cdot m_0(p_i).$$

Example 2.26 *The homogeneous system of equations*

$$\begin{pmatrix} y_1 \\ y_2 \\ y_3 \\ y_4 \end{pmatrix}^T \cdot \mathcal{C}_{\mathcal{N}_1} = 0 \ \ in \ \mathbb{N}$$

has the solutions

$$y_1 = y_2 = y_4 = s \ \ for \ \ s \in \mathbb{N} \ and \ y_3 = 0.$$

Therefore, the following equality holds for each marking $m \in \mathcal{R}_{\mathcal{N}_1}$:

$$s \cdot m(P_1) + s \cdot m(P_2) + 0 \cdot m(P_3) + s \cdot m(P_4) =$$
$$s \cdot m_0(P_1) + s \cdot m_0(P_2) + 0 \cdot m_0(P_3) + s \cdot m_0(P_4)$$

i.e.,

$$m(P_1) + m(P_2) + m(P_4) = 1.$$

All the properties introduced here are qualitative properties, even though some of them might also be considered as quantitative properties. The question of boundedness of a place for instance is equivalent to the question of the maximum number of tokens on the place, which deals with quantities. We will consider quantitative properties that do not have direct qualitative equivalents for the time-dependent Petri nets presented in the following chapters.

2.5 Bibliographical Notes

With his thesis "Kommunikation mit Automaten"[5] [Pet62] Petri established a theory of communication in arbitrarily large, non-globally defined systems. The starting point of his considerations is the opinion that the notion *state* in the sense of one global state is unsuitable for the description of causal relationships. The reason for this is that the use of a notion of global state

---

[5]"Communication with Automata". Published English translation in: Technical Report RADC-TR-65-377, Vol.1, January 1966, Suppl. 1, Griffiss Air Force Base, New York.

presupposes an explicit or implicit time scale implying simultaneity of independent events about whose independence we do not necessarily want to make any presumptions. Petri uses the notions *condition* and *event* for the description of causal relations and thereby defines action nets (or A-nets). Further works by him, as well as Genrich [Gen68], Lautenbach [Lau73], Holt et al. [CHEP71], Commoner [Com73], Hack [Hac72], etc. form the foundation of net theory. This net theory is today known as the theory of Petri nets. The graphic representation of these nets as well as their name "Petri nets" goes back to A. Holt. Innumerable introductory articles and books on the theory of Petri nets have been published since then. Among the first ones and still standard references in the field are the articles by Peterson [Pet77] and Murata [Mur89] as well as the books by Starke [Sta80] and Peterson [Pet81].

The latest book publications include a book by Reisig [Rei13] in which the author presents a thorough introduction to the essentials of Petri nets.

Mayr set another milestone for Petri net theory with his Ph.D. Thesis (cf. [May80]). Among other things he proved the decidability of the reachability of an arbitrary marking in a Petri net.

Algorithms deciding properties for different restricted classes of Petri nets can be found in [Esp98]. Priese and Wimmel give an almost complete introduction to the theory of Petri nets with their book [PW03].

2.6   Exercises

Exercise 2.1

Let $\mathcal{N}_1 = (P, T, F, V, m_0)$ be the following Petri net:

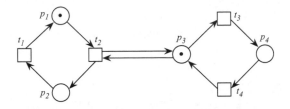

Figure 2.4:   The Petri net $\mathcal{N}_1$

(a) Give the incidence matrix $C_{\mathcal{N}_1}$.

(b) Give the Parikh vectors of the two transition sequences $\sigma_1 = t_3 t_4 t_3 t_4 t_2 t_3$ and $\sigma_2 = t_1 t_3 t_4 t_3^3 t_2 t_1$. Compute the markings m_i which are reached after the firing of the transition sequences $\sigma_i, i = 1, 2$, in the Petri net $\mathcal{N}_1$ starting at m_0 by means of its state equations. What can you say about the reachability of m_1 and m_2 in $\mathcal{N}_1$?

(c) Compute the P- and T-invariants for the Petri net $\mathcal{N}_1$ if it has any. Give a feasible T-invariant if there is one.

(d) Compute the reachability graph of $\mathcal{N}_1$. Is $\mathcal{N}_1$ live?

Exercise 2.2

Let $\mathcal{N}_2 = (P, T, F, V, m_0)$ be the following Petri net:

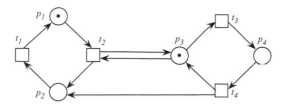

Figure 2.5: The Petri net $\mathcal{N}_2$

(a) Show by means of the state equation that the empty marking $(0, 0, 0, 0)$ is not reachable in $\mathcal{N}_2$.

(b) Compute the coverability graph of $\mathcal{N}_2$. Is $\mathcal{N}_2$ bounded?

Exercise 2.3

(a) Is it always possible to show the non-reachability of an arbitrary marking in a Petri net using the state equation? Give a proof of your answer.

(b) Is the liveness of an unbounded Petri net decidable from its coverability graph?

Exercise 2.4

Let $\mathcal{PN}$ be the set of all unmarked Petri nets and let $\mathcal{M}$ be the set of all (finite) matrices over the natural numbers.

(a) Is the following mapping Φ bijective:

$$\Phi : \mathcal{PN} \longrightarrow \mathcal{M}, \text{ where } \Phi(\mathcal{N}) := C_{\mathcal{N}}?$$

Justify your answer.

(b) If Φ is not bijective in general then give, if possible, a subset of $\mathcal{PN}$ such that the mapping Φ is bijective on this subset. Justify your answer.

Exercise 2.5

An arbitrary Petri net property α is called *monotone* if it holds that: For any Petri net $\mathcal{N} = (P, T, F, V, m_0)$ with the property α, α holds for all Petri nets $\mathcal{N}_k := (P, T, F, V, m_0^k)$ with $m_0^k(p) := k \cdot m_0(p)$ for all $p \in P$.

Is liveness a monotone property? Give a proof.

Chapter 3

Time Petri Nets

In this chapter we study Time Petri nets: After the introduction of this kind of time-dependent Petri net, we will discuss variations of the rules defining the possible state changes. We then show how to reduce the state space of an arbitrary Time Petri net to a discrete one and use this to study the dynamic behavior of these time-dependent Petri nets under both qualitative and quantitative aspects.

3.1 Basic Notions

Time Petri nets (TPN) are classic Petri nets where each transition t is associated with a time interval $[a_t, b_t]$. When t becomes enabled, it cannot fire before a_t time units have elapsed, and it has to fire no later than b_t time units after being enabled, unless it has meanwhile become disabled by the firing of another transition. Here a_t and b_t are relative to the point in time when t last became enabled. The time a_t is the earliest possible firing time for t and is called *earliest firing time* of t (short: $eft(t)$), and b_t is the latest possible firing time for t and is called *latest firing time* of t (short: $lft(t)$). The firing of a transition itself does not take up any time. The interval bounds are non-negative rational numbers or ∞ in the case of b_t but the time interval itself is given in real numbers. In the following, we see (cf. [Pop91], too) that without loss of generality we can require the interval bounds to be natural numbers. Thus, we consider as interval bounds a_t and b_t of a transition t non-negative integers including zero such that $a_t \leq b_t$ or $b_t = \infty$.

Definition 3.1 (Time Petri net) *A Time Petri net (TPN) is a 6-tuple* $\mathcal{Z} = (P, T, F, V, m_0, I)$ *such that*

1. *the 5-tuple* $S(\mathcal{Z}) = (P, T, F, V, m_0)$ *is a Petri net,*

2. $I : T \longrightarrow \mathbb{Q}_0^+ \times (\mathbb{Q}_0^+ \cup \{\infty\})$ *and for each* $t \in T$, *with* $I(t) = (\, I_1(t)\, ,\, I_2(t)\,)$ *it holds that* $I_1(t) \leq I_2(t)$.

Example 3.2

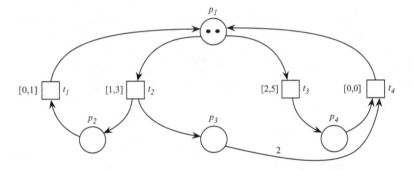

Figure 3.1: The Time Petri net $\mathcal{Z}_1$

The classic (timeless) Petri net $S(\mathcal{Z})$ is called the *skeleton* of $\mathcal{Z}$. I is the *interval function* of $\mathcal{Z}$, $I_1(t)$ and $I_2(t)$ the *earliest firing time of* t and the *latest firing time of* t (short: $eft(t)$ and $lft(t)$), respectively.

We achieve integer interval bounds by scaling down the time unit. Let us, for instance, consider Fig. 3.2. The interval bounds a and b have values 1.5 and 3.0 in the red scale, so a is not an integer. The values of the same bounds in the green scale on the other hand are 3.0 and 6.0, that is, both are non-negative integers. One red time unit is twice as long as one green time unit so the red time unit has been scaled down by a factor of two in order to conform to the green time unit.

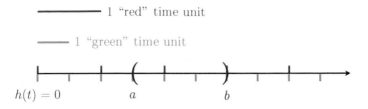

Figure 3.2: Two time scales: The green one is a scaling down of the red one (1 red time unit = 2 green time units).

The smallest suitable factor to scale down the time unit for a given Time Petri net is the lowest common multiple (LCM) of the denominators of all interval bounds in the considered net (∞ excluded). We compute the values of the interval bounds w.r.t. the new time unit by multiplying the old values by this factor. Using the LCM method, we obviously obtain as new values only non-negative integers or ∞. Thus, as of now and without loss of generality, we consider the set $\mathbb{N} \times (\mathbb{N} \cup \{\infty\})$ as codomain for the interval function I. In other words, we use only non-negative integers or ∞ as interval bounds.

Adding time to a Petri net changes the meaning of a marking for the net: a marking no longer describes the full current situation in the time-dependent net. In addition to the number of tokens on each place, i.e., the marking, we now also need to take into account for each enabled transition the amount of time that has passed since its last enabling. We therefore consider a further marking associating with each enabled transition this time and the symbol $\natural$ with each disabled transition. This marking is a vector of dimension equal to the number of transitions. Thus, the first marking describes the situation of the places and the second one the situation of the transitions. We call them the *place-marking* (short: p-marking) and *transition-marking* (short: t-marking), respectively. A p-marking and a t-marking together fully describe the situation in a Time Petri net. Such a pair (p-marking , t-marking), called a *state*, is one of the basic notions in the theory of time-dependent Petri nets. The definition is as follows:

Definition 3.3 (p-marking) *Let P be the set of all places in a Time Petri net $\mathcal{Z}$. A p-marking in $\mathcal{Z}$ is a (total) function $m : P \longrightarrow \mathbb{N}$.*

It is obvious that any p-marking in a Time Petri net $\mathcal{Z}$ is also a marking in its skeleton, the classic Petri net $S(\mathcal{Z})$.

Definition 3.4 (t-marking) *Let T be the set of all transitions in a Time Petri net $\mathcal{Z}$. Any (total) function $h : T \longrightarrow \mathbb{R}_0^+ \cup \{\sharp\}$ is a t-marking in $\mathcal{Z}$.*

Definition 3.5 (state) *Let $\mathcal{Z} = (P, T, F, V, m_o, I)$ be a Time Petri net, m a p-marking and h a t-marking in $\mathcal{Z}$. A state in $\mathcal{Z}$ is a pair $z := (m, h)$ such that*

    1. $\forall t \, (\, (t \in T \wedge t^- \not\leq m) \; \longrightarrow \; h(t) = \sharp).$

    2. $\forall t \, (\, (t \in T \wedge t^- \leq m) \; \longrightarrow \; (h(t) \in \mathbb{R}_0^+ \wedge h(t) \leq lft(t)) \,).$

Evidently, not every pair (m, h) of a p-marking m and a t-marking h is a state in a Time Petri net. The p-marking and the t-marking also need to be suitable for each other. This is the case if for each t the time $h(t)$ is a number (and not $\sharp$) if and only if t is enabled in m. Additionally, we consider the time $h(t)$ of each transition t only until the latest possible firing time for t. This is obviously not a loss of generality.

The state $z_o := (m_o, h_o)$ with $h_o(t) := \begin{cases} 0 & \text{if} \quad t^- \leq m_0 \\ \sharp & \text{if} \quad t^- \not\leq m_0 \end{cases}$ is called *the initial state* of the Time Petri net $\mathcal{Z} = (P, T, F, V, m_0, I)$.

We could of course use other suitable t-markings for the initial state. We will see later on that any vector of rational numbers which is a suitable t-marking for the p-marking m_o can be used as h_0. Such initial states are necessary for modeling and analyzing biochemical systems.

Up until now, we have introduced the static aspects of a Time Petri net. As with classic Petri nets the dynamic aspect is determined by the firing rule(s). The current situation of a Time Petri net may change due to changes in the current p-marking or the t-marking. As for markings in classic Petri nets a change of the p-marking occurs by firing of transitions. Such a firing generally not only changes the current p-marking but also the t-marking. The t-marking however also changes with the elapsing of time even without any transitions firing.

Before we give the definition of firing rules for Time Petri nets we introduce the notion of *a transition being ready to fire*. Where we before only needed to know which transitions were enabled we now distinguish between enabled transitions that have already reached their earliest firing time and enabled transitions for which this is not the case.

In the next three definitions we consider an arbitrary Time Petri net $\mathcal{Z} = (P, T, F, V, m_0, I)$.

Definition 3.6 (ready to fire) *A transition t in $\mathcal{Z}$ is ready to fire in the state $z = (m, h)$ if*

    1. t is enabled in the marking m in the Petri net $S(\mathcal{Z})$, i.e., $t^- \leq m$, and

    2. $h(t) \geq eft(t)$.

1st rule for state change:

Definition 3.7 (firing) *Let $\hat{t}$ be a transition and $z - (m, h)$ be a state in $\mathcal{Z}$. Then $\hat{t}$ can fire in z if $\hat{t}$ is ready to fire in z (notation: $z \xrightarrow{\hat{t}}$). After the firing of $\hat{t}$ the net $\mathcal{Z}$ changes from z into the state $z' = (m', h')$ (notation: $z \xrightarrow{\hat{t}} z'$) with*

    1. $m' := m + \Delta\hat{t}$,

$$2.\ \forall t\ (\ t \in T \longrightarrow h'(t) := \begin{cases} \sharp & \text{if} & t^- \not\leq m' \\ h(t) & \text{if} & t^- \leq m \wedge t^- \leq m' \wedge \\ & & {}^\bullet t \cap {}^\bullet \hat{t} = \emptyset \wedge t \neq \hat{t} \\ 0 & \text{otherwise} \end{cases}\ ).$$

We denote the change from the state z into the state z' by firing of the transition $\hat{t}$ by $z \xrightarrow{\hat{t}} z'$.

Thus, with the firing rule the following has been determined:

- If, after the firing of one of two enabled transitions with at least one common pre-place the second transition is still enabled, its clock is reset to zero.

The rule can, of course, be changed so that in situations such as above we let the transitions stay enabled without resetting their clocks.

A further possibility to define the firing rule is to allow self-concurrency, i.e., a transition is enabled severalfold in a state if its pre-places hold enough tokens.

- The firing itself takes no time. At first glance this seems to be a restriction but actually it is not. As a matter of fact, we could additionally assign a certain time to each transition but this form of time-dependent Petri nets can be simulated by our Time Petri nets. This will be discussed in the next chapter.

2nd rule for state change:

Definition 3.8 (elapsing of time) *Let τ be a non-negative real number and $z = (m, h)$ a state in $\mathcal{Z}$. Then the elapsing of time τ starting at z is possible (notation: $z \overset{\tau}{\longrightarrow}$), if*

$$\forall t \; (\; (t \in T \wedge h(t) \neq \natural) \; \longrightarrow h(t) + \tau \leq lft(t) \;).$$

After τ time has elapsed $\mathcal{Z}$ changes from z into the state $z' = (m', h')$ (notation: $z \overset{\tau}{\longrightarrow} z'$) with

1. *$m' := m$,*

2. *$\forall t \; (\; t \in T \longrightarrow h'(t) := \begin{cases} h(t) + \tau & \text{iff} \quad t^- \leq m' \\ \natural & \text{iff} \quad t^- \not\leq m' \end{cases} \;).$*

This second rule for state change implies that time always passes at equal pace for all enabled transitions. However this does not mean that time always has to pass equally quickly.

Informally, every Time Petri net can be understood as a classic Petri net where each transition has a clock. For a disabled transition the clock stops, but as soon as the transition is enabled its clock starts measuring the time so that for every enabled transition its clock shows the time elapsed since the transition last became enabled. Considering a state $z = (m, h)$ we can interpret $h(t)$ as the clock of t. When $h(t)$ has the value $\natural$ we say that the clock of t has stopped. Any clock of a transition t with $h(t) \in \mathbb{R}_0^+$ is a running clock and shows the time of t in state z.

At the end of this section we note that classic Petri nets can be understood as Time Petri nets where each transition t is assigned the interval $[0, \infty]$. The lower bound $eft(t) = 0$ allows t to fire as soon as enabled and the upper bound $lft(t) = \infty$ indicates that firing of t is not enforced at any point of time.

3.2 State Space

In the following we will introduce some notations enabling us to define the basic notions of a *reachable state* and the *state space of a Time Petri net* analogously to and consistently with the respective notions for classic Petri nets.

Let $\mathcal{Z} = (P, T, F.V, m_o, I)$ be an arbitrary Time Petri net and let $\sigma = t_1 \cdots t_n$ be a transition sequence in T. Whether σ is a firing sequence depends not only on the enabling of the transitions. Every transition t must also have been enabled long enough (be "old enough"), it must have been last enabled for at least $eft(t)$ time units. This in turn means a certain amount of time in general must have passed between the firing of two transitions. Let $\tau = \tau_0 \tau_1 \ldots \tau_n$ with $\tau_i \in \mathbb{R}_0^+$ be a sequence of times. Then the sequence $\sigma(\tau) = \tau_0 t_1 \tau_1 \cdots t_n \tau_n$ is called a *run* of σ.

Definition 3.9 (feasible run) *Let $\mathcal{Z} = (P, T, F, V, m_o, I)$ be a Time Petri net, $z = (m, h)$ a state in $\mathcal{Z}$ and $\sigma(\tau) = \tau_0 t_1 \tau_1 t_2 \cdots t_n \tau_n$ a run of σ. $\sigma(\tau)$ fires from z into z', (short: $z \xrightarrow{\sigma(\tau)} z'$), if*

Basis: *For $n = 0$, i.e., $\sigma = \varepsilon$*

$z' := z$

Step: *For $n \longrightarrow n+1$, i.e., $\sigma = \tau_0 t_1 \cdots t_n \tau_n t_{n+1} \tau_{n+1}$*

There are states z^* and z^{**} in $\mathcal{Z}$ for which it holds:

$$z \xrightarrow{\tau_0 t_1 \tau_1 \cdots t_n \tau_n} z^* \quad and \quad z^* \xrightarrow{t_{n+1}} z^{**} \quad and \quad z^{**} \xrightarrow{\tau_{n+1}} z'.$$

The run $\sigma(\tau)$ is called feasible from the state z in $\mathcal{Z}$, if there is a state z' such that $\sigma(\tau)$ can fire from z into z'.

The run $\sigma(\tau)$ is called feasible in $\mathcal{Z}$, if $\sigma(\tau)$ is a feasible run from z_0 in $\mathcal{Z}$.

It follows immediately from the definition that for any feasible run $\sigma(\tau)$ in a Time Petri net $\mathcal{Z}$ the sequence σ is a firing sequence in the skeleton $S(\mathcal{Z})$ of $\mathcal{Z}$.

Definition 3.10 (firing sequence) *A transition sequence σ is a firing transition sequence in the Time Petri net $\mathcal{Z}$ if there is a feasible run $\sigma(\tau)$ in $\mathcal{Z}$.*

Definition 3.11 (reachable state) *A state z is called reachable in the Time Petri net $\mathcal{Z}$ if there exists a firing sequence σ in $\mathcal{Z}$ with $z_0 \xrightarrow{\sigma} z$.*

Definition 3.12 (state space) *The set $RS_{\mathcal{Z}}$ of all reachable states in a Time Peri net $\mathcal{Z}$ is called the state space of $\mathcal{Z}$.*

Definition 3.13 (reachable p-marking) *A p-marking m is called reachable in a Time Petri net $\mathcal{Z}$ if there is a reachable state z in $\mathcal{Z}$ with $z = (m, h)$.*

We use the notation $R_{\mathcal{Z}}$ for the *set of all p-markings reachable in $\mathcal{Z}$*. A Time Petri net $\mathcal{Z}$ is *bounded* if $R_{\mathcal{Z}}$ is finite.

Adding explicit time to Petri nets (in this way) in general restricts the behavior possible in a Time Petri net compared to its skeleton. The set of all reachable p-markings in the time-dependent net is therefore a subset of the state space of its skeleton, i.e., $R_{\mathcal{Z}} \subseteq R_{S(\mathcal{Z})}$. This is a consequence of *forcing* any transition that is still enabled at the end of its time interval to fire.

In the case of all transitions being associated with the interval $[0, \infty]$ the set of all reachable p-markings of the Time Petri net is equal to the state space of its skeleton, i.e., $R_{\mathcal{Z}} = R_{S(\mathcal{Z})}$. We will see later that whereas this is a sufficient condition for the equality of these sets it is not a necessary one.

3.3   TPN-Computability

We saw in Chapter 2 that classical Petri nets do not have the same computational power as Turing machines, i.e., that there exist algorithms which cannot be simulated (modeled) using classic Petri nets.

Adding time to classic Petri nets and modifying the firing rule we have introduced Time Petri nets. We will show in this section that Time Petri nets have the same computational power as Turing machines. We will prove this by verifying that each number-theoretical function which is computable by a counter machine it is also computable by a Time Petri net. Because counter machines (with at least two counters) have the same computational power as Turing machines (cf. [HMU02]), the equivalence of the computational power of Time Petri nets and Turing machines follows.

Informally, a counter machine is a restricted multistack machine which can store a finite number of natural numbers, and can add or subtract (if the counter is not already zero) one to or from any of these counters. For more on multistack machines, see [HMU02].

We now formally introduce the counter machine. This definition can also be found in [Sta80]. A *counter machine* consists of finitely many counters $K = \{K_1, \cdots, K_k\}$ and a program. Each counter can store one natural number[1]. The program is a finite uniquely numbered list of commands. The available commands are: *start, halt, INC* and *DEC*, the so-called zero-test. For a labeled command we use the notation *assignment*. In each program the command *start* appears exactly once[2] and *halt* at least once. The notation and mode of operation of the four commands is described in Table 3.1.

Notation of the command	mode of operation of the command
0 : start : l	start the program; go to command No. l ;
l : halt	stop the program;
l : INC(i) : r	$K_i := K_i + 1$; go to command No. r ;
l : DEC(i) : r : s	if $K_i = 0$ then go to command No. r ; else $K_i := K_i - 1$; go to command No. s ;

Table 3.1: The four possible commands of a counter machine with their meaning

---

[1]We often identify the number stored in a counter with the counter, i.e., we say "the number K_i" instead of "the number stored in K_i".

[2]Without loss of generality we assume that the command *start* occures exactly once during any execution of a program.

The first number in each assignment is its unique label in the program. Each of the four commands can be simulated by a (small) Time Petri net. Starke shows in [Sta80] that the first three commands can be simulated by classic Petri nets. Thus, for the fourth one (DEC) this is not possible. We follow this general approach, adapt it to Time Petri nets, and complete it through a simulation of the command DEC by a Time Petri net: Each number l of an assignment is modeled by a place p_l such that a place holds a token whenever the corresponding assignment is being executed. Furthermore, each counter K_i is modeled by a place w_i. During the whole computation process (program execution) it will always hold that $m(w_i) = K_i$ for the p-marking m in the net corresponding to the current step in the computation.

We model the commands through Time Petri nets as indicated in Table 3.2.

The intervals in the second and third models have been chosen arbitrarily. We can use any valid interval in these nets.

In the model of the fourth command (DEC) it is important to prioritize transition t_2 using intervals. We ensure that any conflict between the transitions t_1 and t_2 is decided in favor of t_2 by letting $lft(t_2) < eft(t_1)$.

Definition 3.14 (CM-computable) *An n-ary number-theoretical function f is called counter-machine-computable (short: CM-computable) if there is a counter machine with k counters $\{K_1, \cdots, K_k\}$ where $k \geq n$ and for each n-tuple $(x_1, \cdots, x_n) \in \mathbb{M}^n$ it holds that for the counter machine started with the values*

$$K_i = \begin{cases} x_i & \text{if } 1 \leq i \leq n \\ 0 & \text{if } n < i \leq k \end{cases},$$

in the counters:

Case 1 *if $(x_1, \cdots, x_n)$ belongs to the domain of the function f the machine must reach a command 'halt' after a finite number of steps, i.e., the computation must terminate, and on termination the number $f(x_1, \cdots, x_n)$ is stored in the first counter K_1;*

and

Notation of the command	Model of the command as a Time Petri net
$0 : start : l$	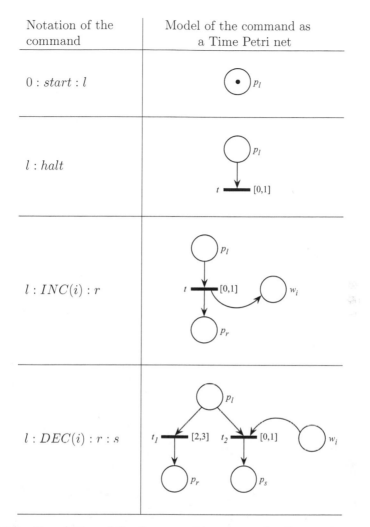
$l : halt$	
$l : INC(i) : r$	
$l : DEC(i) : r : s$	

Table 3.2: Translation of the four possible commands of a counter machine into Time Peri nets (modules)

Case 2 *if $(x_1, \cdots, x_n)$ does not belong to the domain of the function f the machine never reaches a command 'halt', i.e., the computation does not terminate.*

TPN-computability is defined similarly to PN-computability:

Definition 3.15 (TPN-computable) *An n-ary function f is called Time-Petri-net-computable (TPN-computable) if there is an initial Time Petri net $\mathcal{Z}_f = (P_f, T_f, F_f, V_f, I_f, m_0^f)$ such that for every n-tuple $x = (x_1, \cdots, x_n) \in \mathbb{N}^n$ and for the Time Petri net $\mathcal{Z}_f^x = (P_f, T_f, F_f, V_f, I_f, m_0^{f,x})$, where the p-marking $m_0^{f,x}$ models the n-tuple $(x_1, \cdots, x_n)$,[3] it holds that:*

Case 1 *If the tuple $(x_1, \cdots, x_n)$ belongs to the domain of f then the Time Petri net $\mathcal{Z}_f^x$ stops (cannot fire anymore) and the last-reached p-marking $m^{f(x_1, \cdots, x_n)}$ uniquely models the number $f(x_1, \cdots, x_n)$.*

Case 2 *If the tuple $(x_1, \cdots, x_n)$ does not belong to the domain of f then the Time Petri net $\mathcal{Z}_f^x$ never stops, i.e., for each state z with $z_0^{f,x} \xrightarrow{*} z$ there exist at least one non-negative real number τ and a transition t such that $z \xrightarrow{\tau} \xrightarrow{t}$.*

It is evident that every CM-computable function is also TPN-computable, and vice versa. We refrain from proving this formally. The idea is now to let a Time Petri net simulate the program of a counter machine. In this net no transition is ready to fire unless at least one place p_i is marked. In each reachable p-marking exactly one place p_i is marked but in the marking reached after firing a transition corresponding to a command $l : halt$, no place p_i is marked anymore.

Thus, we can summarize:

Proposition 3.16 *Let f be an n-ary number-theoretical function. Then it holds that: f is Turing-computable if and only if f is TPN-computable.*

This follows immediately from the above argument and the equivalence of Turing machines and counter machines.

In the next two Examples 3.17 and 3.18 we consider one total function and a partial one in order to illustrate the representation of an arbitrary number-theoretical function with a Time Petri net. To this end we translate the program of a counter machine which computes the respective function into a Time Petri net as indicated in Table 3.2.

---

[3]The p-marking $m_0^{f,x}$ takes into account the initial p-marking m_0^f.

Example 3.17 *Let us first consider the addition of two natural numbers, which is a 2-ary total function. The initial Time Petri net $\mathcal{Z}_f$ is constructed using the program of the counter machine which computes the function. For the p-marking m_0^f it holds that:*

$$m_0^f(p_1) = 1, \quad m_0^f(p_2) = m_0^f(p_3) = 0 = m_0^f(w_1) = m_0^f(w_2).$$

$f(x_1, x_2) = x_1 + x_2$	
Program of the counter machine	*Time Petri net model*
0 : *start* : 1 1 : $DEC(2)$: 3 : 2 2 : $INC(1)$: 1 3 : *halt*	

Consider the pair $x = (x_1, x_2) = (2, 3)$. The Time Petri net $\mathcal{Z}_f^{(2,3)}$ which computes the function f for the argument $(2,3)$ is obtained from the initial Time Petri net $\mathcal{Z}_f$ by adding tokens to some places in the initial p-marking m_0^f. The initial p-marking m_0^f is extended to $m_0^{f,(2,3)}$ as follows:

$$m_0^{f,(2,3)}(p_i) = m_0^f(p_i) \text{ for } i = 1, 2, 3, \quad m_0^{f,(2,3)}(w_1) = 2, \; m_0^{f,(2,3)}(w_2) = 3.$$

In the net $\mathcal{Z}_f^{(2,3)}$ only the transition sequence $t_2 t_3 t_2 t_3 t_2 t_3 t_1 t_4$ can fire. After that no transition is enabled any more, i.e., the net stops in the state $z = (m, h)$ with $m(w_1) = m_0(w_1) + m_0(w_2)$. Thus the place w_1 contains five tokens, which is equal to the value $f(2, 3)$.

Example 3.18 *As a further example, let us consider the subtraction of two natural numbers, which is a 2-ary partial function. It is formally defined as follows:*

$$g(x_1, x_2) = \begin{cases} x_1 - x_2 & \text{if } x_1 \geq x_2 \\ \text{undefined} & \text{otherwise} \end{cases}.$$

We deduce the initial Time Petri net $\mathcal{Z}_g$ from the counter machine program computing the function g as before:

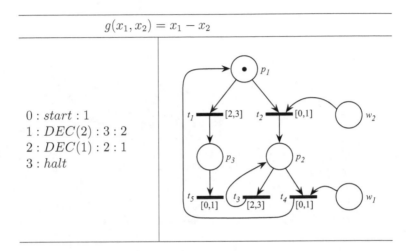

$g(x_1, x_2) = x_1 - x_2$

$0 : start : 1$
$1 : DEC(2) : 3 : 2$
$2 : DEC(1) : 2 : 1$
$3 : halt$

As before, a Time Petri net $\mathcal{Z}_g^x$ for an arbitrary pair $x = (x_1, x_2)$ of arguments for g is obtained from the initial net $\mathcal{Z}_g$ by adding tokens to the initial p-marking m_0^g. We now consider the pair $(2,3)$. It is obvious that the function g is undefined for those arguments. Let us consider the initial p-marking $m_0^{g,(2,3)}$ arising from m_0^g:

$$m_0^{g,(2,3)}(p_i) = m_0^g(p_i) \text{ for } i = 1, 2, 3, \quad m_0^{g,(2,3)}(w_1) = 2, \ m_0^{g,(2,3)}(w_2) = 3.$$

It is easy to see that for an arbitrary pair $x = (x_1, x_2)$ the Time Petri net $\mathcal{Z}_g^x$ stops if and only if $m_0(w_1) \geq m_0(w_2)$. In particular, the net $\mathcal{Z}_g^{(2,3)}$ never stops – first the transition sequence $t_2 t_4 t_2 t_4 t_2$ fires and after that the transition t_3 can fire infinitely often.

3.4 State Space Reduction

The state space of a Petri net, timeless or time-dependent, contains total knowledge of its behavior. Explicit knowledge of the state space in general is indispensable for studying the dynamic properties of a net but as the state space of an arbitrary net is usually infinite we will show in this section how to reduce the state space without affecting the properties of the net.

We first give a parametric description of the state space of Time Petri nets and then reduce it. We will subsequently prove necessary and sufficient conditions under which an infinite state space can be reduced to a finite one. The reduced state space will be well suited for model checking. It will furthermore permit quantitative analysis of a net, which is otherwise only partly possible.

The reduced state space will also enable us to define the *reachability graph of a Time Petri net*, which we examine in the next section. Eventually, we will present some efficient, graph-theoretical algorithms for the analysis of finite reachability graphs.

Let us now consider an arbitrary Time Petri net $\mathcal{Z} = (P, T, F, V, m_0, I)$. Furthermore let $\sigma = t_1 \cdots t_n$ be a firing sequence in $\mathcal{Z}$ and let there be at least one feasible run $\sigma(\tau) = \tau_0 t_1 \tau_1 \cdots \tau_{n-1} t_n \tau_n$ of σ in $\mathcal{Z}$, i.e., there is a state z^* in $\mathcal{Z}$ with

$$z_0 \xrightarrow{\sigma(\tau)} z^* = (m^*, h^*) \qquad \text{and} \qquad m^* = m_0 + \sum_{i=1}^{n} \Delta t_i.$$

It is obvious that the p-marking m^* does not depend on τ, meaning that after the firing of any feasible run of σ (starting in the same state) the reached p-marking is the same; it depends only on σ. This, however, is not true for the t-marking which depends on σ as well as on τ.

Now, instead of considering infinitely many feasible runs of σ we study a single *parametric run* $\sigma(x) = x_0 t_1 x_1 \cdots x_{n-1} t_n x_n$ of σ. Instead of a fixed number τ_i we use the variable x_i to denote the time elapsing between the firing of the i-th and the $i+1$-th transition in σ and look at the requirements for values $\beta(x_i)$ of x_i $(i = 1, \cdots, n)$ that make the run $\sigma(\beta(x))$ a feasible one. Thus, for the state z_σ, with

$$z_0 \xrightarrow{\sigma(x)} z_\sigma = (m_\sigma, h_\sigma),$$

it holds that $m_\sigma = m^*$.

The conditions for the values $\beta(x_i)$ result from the time intervals associated with the transitions and are united into the set B_σ. The state z_σ together with the set B_σ forms the *parametric state* (z_σ, B_σ). Such a parametric state also represents the set of all states which can be reached by firing a feasible run of σ. These states are obtained by combining σ with all possible solutions x of B_σ. Each such solution yields a feasible run of σ. We denote the set of all states by $\{z_\sigma \mid B_\sigma\}$, i.e.,

$$\{z_\sigma \mid B_\sigma\} := \{z_{\sigma(\beta(x))} \mid \beta(x) \text{ is a solution of } B_\sigma\}.$$

We will now recursively define the *parametric state* and the *parametric run* of a transition sequence.

Definition 3.19 (parametric state and parametric run)
Let $\mathcal{Z} = (P, T, F, V, m_0, I)$ be a Time Petri net and let $\sigma = t_1 \cdots t_n$ be a firing sequence in $\mathcal{Z}$. Then, the parametric run $(\sigma(x), B_\sigma)$ of σ in $\mathcal{Z}$ with $\sigma(x) = x_0 t_1 x_1 \cdots x_{n-1} t_n x_n$ and the parametric state (z_σ, B_σ) in $\mathcal{Z}$ are recursively defined as follows:

Basis: $\sigma = \varepsilon$, , *i.e.,* $\sigma(x) = x_0$.

   Then $z_\sigma = (m_\sigma, h_\sigma)$ and B_σ are defined as follows:

   1. $m_\sigma := m_o$,

   2. $h_\sigma(t) := \begin{cases} x_0 & \text{if } t^- \leq m_\sigma \\ \sharp & \text{otherwise} \end{cases}$,

   3. $B_\sigma := \{\, 0 \leq h_\sigma(t) \leq lft(t) \mid t \in T \wedge t^- \leq m_\sigma \,\}$

Step: *Assume that z_σ and B_σ are already defined for the sequence $\sigma = t_1 \cdots t_n$.*

   For $\sigma = \underbrace{t_1 \cdots t_n}_{:=w} t_{n+1} = w t_{n+1}$ we set

   1. $m_\sigma := m_w + \Delta t_{n+1}$,

$$2. \ h_\sigma(t) := \begin{cases} \sharp & \text{if } t^- \not\leq m_\sigma \\ h_w(t) + x_{n+1} & \text{if } t^- \leq m_\sigma \wedge t^- \leq m_w \wedge \\ & {}^\bullet t_{n+1} \cap {}^\bullet t = \emptyset \wedge t \neq t_{n+1} \\ x_{n+1} & \text{otherwise} \end{cases},$$

3. $B_\sigma := B_w \ \cup \ \{ \, eft(t_{n+1}) \leq h_w(t_{n+1}) \, \} \ \cup \ \{ \, 0 \leq h_\sigma(t) \leq lft(t) \ | \ t \in T \wedge t^- \leq m_\sigma \, \}.$

So the t-marking h_σ is a vector with a sum of variables as each of its components. At most $\ell(\sigma) + 1$ variables can appear in h_σ.

The initial set of conditions B_ε consists of only two inequalities and the number of inequalities in each successive set of conditions increases by twice the number of enabled transitions in the respective p-marking plus 1. In the worst case this results in $2 \cdot |T| + 1$ inequalities, of which however many might be redundant.

For a transition sequence σ of length $\ell(\sigma) = n$ the number of variables in B_σ is $n + 1$. The number of inequalities in B_σ on the one hand is no greater than $(2 \cdot |T| + 1) \cdot n + 2 \cdot |T|$. There are, however, at least $(2 \cdot (|T| - 1)) + n$ redundant inequalities in B_σ. Taking into account Remark 3.23 it is not difficult to see that the number of inequalities is also no greater than $(n + 1) \cdot (n + 2)$. Thus, we can summarize that the number of inequalities in B_σ is at most $\min\{2 \cdot (n \cdot |T| + 1), (n + 1) \cdot (n + 2)\}$.

The upper bound $(n + 1) \cdot (n + 2)$ can be lowered to $(n + 1) \cdot (\frac{n}{2} + 2)$. The proof of this is left to the reader, cf. also Exercise 3.3.

For long transition sequences this representation can become rather large. We actually only use it to verify that the reduction given below fulfills the requirements. In applications of Time Petri nets as well as in their analysis we do not need parametric states as long as the reduced state space is finite. This is mostly the case for real technical systems and metabolic networks etc. In case of an infinite reduced state space parametric states are used to study some quantitative properties of the net.

We note that in Definition 3.19 instead of requiring σ to be a firing sequence we could allow it to be an *arbitrary transition sequence*. If σ is not a firing sequence, then it holds that $\{z_\sigma \mid B_\sigma\} = \emptyset$.

The set L_σ of all solutions for $x := (x_0, \cdots, x_{\ell(\sigma)})$ satisfying the inequalities in B_σ is a polyhedron. The t-marking is a vector of linear functions, say $h_\sigma(t) = f_t(x)$, $x \in L_\sigma$ and $f_t(x)$ is linear.

In the next example we will calculate some parametric states. We use K_σ as a shorthand for $\{z_\sigma \mid B_\sigma\}$.

Example 3.20 (parametric states) *Let us consider the Time Petri net* $\mathcal{Z}_2$.

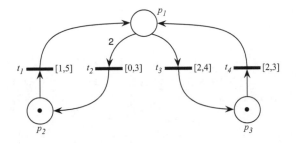

Figure 3.3: The Time Petri net $\mathcal{Z}_2$

It is easy to see that

$$K_\varepsilon = \{(\underbrace{(0,1,1)}_{m_\varepsilon}, \underbrace{(x_0, \sharp, \sharp, x_0)}_{h_\varepsilon}) \mid \underbrace{\{0 \le x_0 \le 3\}}_{B_\varepsilon}\}.$$

After firing the sequence $\sigma = t_4$ *the net* $\mathcal{Z}_2$ *is in a state belonging to* $K_\sigma = K_{t_4}$.

$$K_{t_4} = \{(\underbrace{(1,1,0)}_{m_{t_4}}, \underbrace{(x_0 + x_1, \sharp, x_1, \sharp)}_{h_{t_4}}) \mid \underbrace{\{2 \le x_0 \le 3, x_0 + x_1 \le 5, 0 \le x_1 \le 4\}}_{B_{t_4}}\}.$$

The set of conditions B_{t_4} *is the union of the three sets*

$$B_\varepsilon,$$

$$\{eft(t_4) \le h_\varepsilon(t_4)\} = \{2 \le x_0\} \text{ and}$$

$$\{0 \leq h_\sigma(t) \leq lft(t) \mid t^- \leq m_\sigma\} = \{x_0 + x_1 \leq 5, 0 \leq x_1 \leq 4\}.$$

By repeatedly firing the transitions t_3 and t_4 we obtain the parametric states $z_{t_4 t_3}$ and $z_{t_4 t_3 t_4}$ and $K_{t_4 t_3}$ and $K_{t_4 t_3 t_4}$:

$$K_{t_4 t_3} = \{((0,1,1),(x_0 + x_1 + x_2, \natural, \natural, x_2)) \mid \{ \begin{array}{l} 2 \leq x_0 \leq 3, \quad x_0 + x_1 \leq 5, \\ 2 \leq x_1 \leq 4, \quad x_0 + x_1 + x_2 \leq 5, \\ 0 \leq x_2 \leq 3 \end{array} \} \},$$

$$K_{t_4 t_3 t_4} = \{((1,1,0),(x_0 + x_1 + x_2 + x_3, \natural, x_3, \natural)) \mid \{ \begin{array}{l} 2 \leq x_0 \leq 3, \\ 2 \leq x_1 \leq 4, \\ 2 \leq x_2 \leq 3, \\ 0 \leq x_3 \leq 4, \\ x_0 + x_1 \leq 5, \\ x_0 + x_1 + x_2 \leq 5, \\ x_0 + x_1 + x_2 + x_3 \leq 5 \end{array} \} \}.$$

As mentioned above, parametric states can also be seen as representing sets of reachable states. They can be regarded as dividing the state space into overlapping classes of states. We now define by induction the notion of a *state class* which is also a set of states reachable by firing.

Definition 3.21 (state class)

Let $Z - (P, T, F, V, m_o, I)$ be a TPN and σ a feasible transition sequence. The state class C_σ of σ is defined as follows:

Basis: $C_\varepsilon := \{z \mid \exists \tau (\tau \in \mathbb{R}_0^+ \wedge z_0 \xrightarrow{\tau} z)\}$

Step: If C_σ is already defined then $C_{\sigma t}$ is derived from C_σ by firing t

(denoted by $C_\sigma \xrightarrow{t} C_{\sigma t}$) as follows:

$$C_{\sigma t} := \{z \mid \exists z_1 \exists z_2 \exists \tau (z_1 \in C_\sigma \wedge \tau \in \mathbb{R}_0^+ \wedge z_1 \xrightarrow{t} z_2 \xrightarrow{\tau} z)\}.$$

So the state class C_ε is the set of all states in Z reachable from the initial state by elapsing of time but without firing transitions. For transition sequences $\sigma \neq \varepsilon$ the class C_σ contains all states that are reachable by firing any feasible runs of σ.

The next corollary directly follows from Definitions 2.11, 3.19 and 3.21:

Corollary 3.22 *For every Time Petri net $\mathcal{Z}$ and for every firing sequence σ it holds that:*

(i) $RS_{\mathcal{Z}} = \bigcup_{\sigma} C_\sigma$, *i.e., the state space of $\mathcal{Z}$ is the union of all state classes.*

(ii) $\{z_\sigma \mid B_\sigma\} = C_\sigma$.

The corollary also holds for every transition sequence in $\mathcal{Z}$. As noted before, the set $\{z_\sigma \mid B_\sigma\} = C_\sigma$ is empty for any transition sequence which is not a firing sequence in $\mathcal{Z}$.

Let (z_σ, B_σ) be a parametric state. We want to examine the structure of the inequalities in the set of conditions B_σ and the structure of the linear functions (sums) $h_\sigma(t)$. Each set of conditions is also a system of linear inequalities where all coefficients of variables are 0 or 1. For the sake of simplicity, we will say that *a variable appears* in an inequality or a sum, if its coefficient is 1.

Remark 3.23 *Let $\mathcal{Z}$ be a Time Petri net, σ a firing sequence in $\mathcal{Z}$ and (z_σ, B_σ) with $z_\sigma = (m_\sigma, h_\sigma)$ a parametric state reached after firing σ. Furthermore let $\ell(\sigma) = n$ and $x = (x_0, \cdots, x_n)$. Then it holds that:*

(i) *If t is enabled in m_σ then the variable x_n appears in $h_\sigma(t)$.*

(ii) *If t is enabled in m_σ and $h_\sigma(t) = x_i + \cdots + x_j$ with $i \leq j$ then every variable x_k with $i \leq k \leq j$ appears in the sum $h_\sigma(t)$.*

(iii) *If the transitions t_1 and t_2 are enabled in m_σ then either each variable appearing in $h_\sigma(t_1)$ also appears in $h_\sigma(t_2)$ or vice versa.*

(iv) *If $g(x) \leq r$ is an inequality in B_σ and the variables x_i and x_j with $i \leq j$ appear in $g(x)$ then every variable x_k with $i \leq k \leq j$ also appears in $g(x)$.*

Proof:

(i) The claim follows immediately from Definition 3.19.

(ii) This property can easily be proved by induction on $j - i$.

So, according to *(i)* $h_\sigma(t)$ has the form:

$$
\begin{aligned}
h_\sigma(t) &= x_{n-k} + x_{n-(k-1)} + \cdots + x_{n-(k-k)} \\
&= x_{n-k} + x_{n-(k-1)} + \cdots + x_n
\end{aligned}
$$

for some $k \in \{0, 1, \cdots, n\}$.

(iii) Now, for transitions t_1 and t_2 enabled in m_σ (i) and (ii) lead to:

$$
\begin{aligned}
h_\sigma(t_1) &= x_{n-k} + x_{n-(k-1)} + \cdots + x_{n-(k-k)} \\
&= x_{n-k} + x_{n-(k-1)} + \cdots + x_n \quad \text{and} \\
h_\sigma(t_2) &= x_{n-l} + x_{n-(l-1)} + \cdots + x_{n-(l-l)} \\
&= x_{n-l} + x_{n-(l-1)} + \cdots + x_n.
\end{aligned}
$$

If $k \leq l$ then all variables appearing in $h_\sigma(t_1)$ also appear in $h_\sigma(t_2)$. If $l \leq k$ then all variables appearing in $h_\sigma(t_2)$ also appear in $h_\sigma(t_1)$.

(iv) Induction on n gives that for each inequality $g(x) \leq r$ in B_σ there exist subsequences σ_1 and σ_2 of σ with $\sigma = \sigma_1\sigma_2$ such that $g(x) = h_{\sigma_1}(t)$ for some transition $t \in T$. Thereafter, taking into account (i), claim (iv) follows immediately.

The following Theorem 3.26 states a fundamental property of Time Petri nets, namely that each p-marking reached with an arbitrary feasible run is also reachable with a feasible run where the values of all elapsed times are natural numbers (integers). For such a run all intermediate states reached during the execution of the run are also "integer-states", i.e., the clock of each enabled transition in each intermediate state shows a natural number. These states play a crucial role in our reduction and the analysis of the state space. We will show that they carry most of the net properties: dynamic as well as static and qualitative as well as quantitative.

Definition 3.24 (integer-state) *In a Time Petri net a state $z = (m, h)$ is called an integer-state if for every transition t enabled in m it holds that $h(t) \in \mathbb{N}$.*

Definition 3.25 (integer-run) *In a Time Petri net a run $\sigma(\tau)$ with $\sigma = t_1 \cdots t_n$ and $\tau = \tau_0 \cdots \tau_n$ is called an integer-run if for every $i = 0, \cdots, n$ it holds that $\tau_i \in \mathbb{N}$.*

The notation $RIS_{\mathcal{Z}}$ stands for the set of all reachable integer-states in $\mathcal{Z}$.

Before we go on we declare the following notations: Let X be a set of variables and let β be an assignment of those variables to non-negative real numbers, i.e., $\beta : X \longrightarrow \mathbb{R}_0^+$. Let $x = (x_1, \cdots, x_n)$ be a vector of variables from X. Then $\beta(x)$ stands for the vector of the non-negative real numbers $(\beta(x_1), \cdots, \beta(x_n))$. Furthermore, we denote the value of the linear function $g(x) = x_{i_1} + \cdots + x_{i_k}$ under the assignment β by $[\![g(x)]\!]_\beta$. Finally, $s(c)$ stands for the variable part of an inequality c. Hence, the inequality $c : g(x) \le k$ is satisfied by β if $[\![s(c)]\!]_\beta \le k$.

Theorem 3.26 (Main theorem) *Let $\mathcal{Z} = (P, T, F, V, m_0, I)$ be a Time Petri net, σ a transition sequence of length n, with $z_0 \xrightarrow{\sigma} (z_\sigma, B_\sigma)$, $z_\sigma = (m_\sigma, h_\sigma)$, $\ell(\sigma) = n$ and $X_\sigma := \{x_0, x_1, \ldots, x_n\}$. Furthermore let $\sigma(\hat{\beta}(x))$ with $\hat{\beta} : X_\sigma \longrightarrow \mathbb{R}_0^+$ be a feasible run of σ. Then there exists an assignment $\beta^* : X_\sigma \longrightarrow \mathbb{N}$ for which the following holds:*

(1) $\sigma(\beta^*(x))$ is also a feasible run of σ in $\mathcal{Z}$,

(2) for each transition t enabled in m_σ it holds that $[\![h_\sigma(t)]\!]_{\beta^*} \le [\![h_\sigma(t)]\!]_{\hat{\beta}}$,

(3) the total duration of the run $\sigma(\beta^*(x))$ is at most the total duration of the run $\sigma(\hat{\beta}(x))$, i.e., $\left[\!\left[\sum_{k=0}^{n} x_k \right]\!\right]_{\beta^*} \le \left[\!\left[\sum_{k=0}^{n} x_k \right]\!\right]_{\hat{\beta}}$.

(1) tells us that the feasible run $\sigma(\beta^*(x))$ will be an integer-run. (2) states that after firing the run $\sigma(\beta^*(x))$ the clock of each enabled transition shows a time that is at most the time shown after firing the run $\sigma(\hat{\beta}(x))$. Finally, (3) says that by firing $\sigma(\beta^*(x))$ the final p-marking will not be reached any later than by firing the original run $\sigma(\hat{\beta}(x))$.

In the proof of (2) we will additionally show that

$$0 \le [\![h_\sigma(t)]\!]_{\hat{\beta}} - [\![h_\sigma(t)]\!]_{\beta^*} < 1.$$

This means that if a state is reachable in a Time Petri net then the corresponding "rounded-down" state is reachable as well.

The assignment β^* described in the theorem is not uniquely determined, but the construction given here will define a unique assignment β^*. Also we can always easily find an assignment β^* such that

$$0 \leq [\![\sum_{k=0}^{n} x_k]\!]_{\hat{\beta}} - [\![\sum_{k=0}^{n} x_k]\!]_{\beta^*} < 1.$$

Therefore we can add the redundant inequality $0 \leq x_0 + x_1 + \ldots + x_n$ to B_σ. This ensures that the total duration of the run $\sigma(\beta^*(x))$ is the rounded-down total duration of the original run $\sigma(\hat{\beta}(x))$. Note that this does not mean that the value of each elapsed time between two successive firings in $\beta^*(x_i)$ is the rounded-down value of the respective time in $\hat{\beta}(x_i)$ – for some i it might hold that $\beta^*(x_i) = \lceil \hat{\beta}(x_i) \rceil$. We recommend considering Exercise 3.4 at this point.

Idea of the proof:
In $n+1$ steps the non-negative integer values $\beta^*(x_0), \beta^*(x_1), \ldots, \beta^*(x_n)$ will explicitly be constructed from the values $\hat{\beta}(x_0), \hat{\beta}(x_1), \ldots, \hat{\beta}(x_n)$. In each step we will define a new intermediate assignment from the previous one by rounding down or up exactly one value $\hat{\beta}(x_i)$.

We first assign a natural number to the variable x_n, then to x_{n-1}, etc. In the last step the value $\hat{\beta}(x_0)$ is rounded.

The algorithm starts (in the second step of the recursive definition) by rounding down $\hat{\beta}(x_n)$. Thereafter, depending on the inequalities in B_σ the algorithm continues rounding the other values down or up.

Definition 3.27 (Construction of β^*) *Let* $X_\sigma := \{x_0, x_1, \ldots, x_n\}$ *be the set of all variables appearing in* B_σ. *We define by induction a finite sequence of assignments* $\beta_i : X_\sigma \longrightarrow \mathbb{R}_0^+$:

Basis: $\beta_0 : X_\sigma \longrightarrow \mathbb{R}_0^+$ *with* $\beta_0(x) := \hat{\beta}(x)$ *for all* $x \in X_\sigma$.

Step: *Assume that* β_{i-1} *is already defined. In the construction of* β_i *we use the following function:*

$$\underline{\beta_i}(x) := \begin{cases} \beta_{i-1}(x) & \text{if } x \neq x_{n-(i-1)} \\ \lfloor \beta_{i-1}(x) \rfloor & \text{otherwise} \end{cases}.$$

We now define $\beta_i : X_\sigma \longrightarrow \mathbb{R}_0^+$ by

$$\beta_i(x) := \begin{cases} \beta_{i-1}(x) & \text{if } x \neq x_{n-i+1} \\ \lfloor \beta_{i-1}(x) \rfloor & \text{if } x = x_{n-i+1} \wedge \\ & \quad \forall c (c \in B_\sigma \rightarrow \lfloor [\![s(c)]\!]_{\beta_0} \rfloor - 1 < [\![s(c)]\!]_{\beta_i}) \\ \lceil \beta_{i-1}(x) \rceil & \text{otherwise} \end{cases}.$$

The construction defines the first assignment β_0 to be the original assignment $\hat{\beta}$, i.e., in the 0-th step it sets $\beta_0 := \hat{\beta}$. The assignment β_1 is then obtained from β_0 by rounding down the value $\beta_0(x_n)$. β_2 is obtained from β_1 by rounding the value $\beta_1(x_{n-1})$. In general, assignment β_i is obtained from β_{i-1} by rounding the value $\beta_{i-1}(x_{n-(i-1)})$. For each i the value of $x_{n-(i-1)}$ is rounded down only if for all inequalities c in B_σ the value of the variable part of the inequality under the assignment β_i , i.e., the value $[\![s(c)]\!]_{\beta_i}$ is larger than $\lfloor [\![s(c)]\!]_{\beta_0} \rfloor - 1$. Otherwise the algorithm sets $\beta_{i-1}(x_{n-(i-1)}) := \lceil \beta_{i-1}(x_{n-(i-1)}) \rceil$. Hence, for every assignment β_i and each inequality c in B_σ it holds that the value $[\![s(c)]\!]_{\beta_i}$ lies within the interval $(\lfloor [\![s(c)]\!]_{\beta_0} \rfloor - 1, \lceil [\![s(c)]\!]_{\beta_0} \rceil + 1)$. Fig. 3.4 illustrates for an arbitrary inequality c in B_σ the position of the five numbers that are of relevance here.

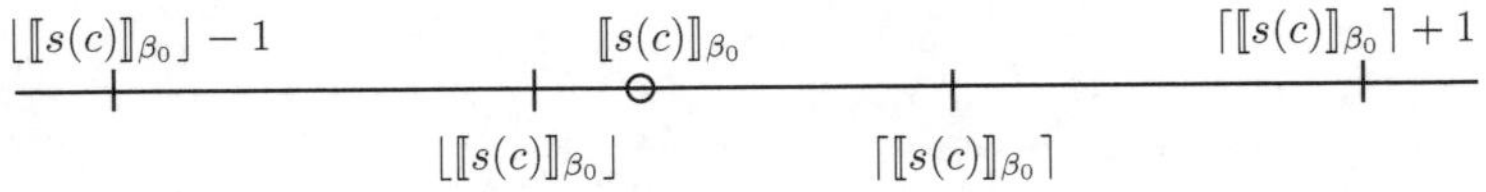

Figure 3.4: Position of the real number $[\![s(c)]\!]_{\beta_0}$ and the integers $\lfloor [\![s(c)]\!]_{\beta_0} \rfloor - 1$, $\lfloor [\![s(c)]\!]_{\beta_0} \rfloor$, $\lceil [\![s(c)]\!]_{\beta_0} \rceil$ and $\lceil [\![s(c)]\!]_{\beta_0} \rceil + 1$.

We can summarize that for all i with $0 \leq i \leq n+1$ and all variables x_k with $0 \leq k < n - (i-1)$ it holds that

$$\beta_i(x_k) = \beta_{i-1}(x_k) = \ldots = \beta_0(x_k) \tag{1}$$

and for all such i and variables x_k with $n \geq k \geq n - (i-1)$

$$\beta_i(x_k) = \beta_{i+1}(x_k) = \ldots = \beta_{n+1}(x_k). \tag{2}$$

Furthermore, if $\beta_0(x_{n-(i-1)})$ is already an integer, then the algorithm leaves the value for $x_{n-(i-1)}$ unaltered in all the assignments, since for any integer k, $k = \lfloor k \rfloor = \lceil k \rceil$.

Obviously the values $\beta_{n+1}(x_0), \beta_{n+1}(x_1), \ldots, \beta_{n+1}(x_n)$ are all integers. Thus, we can define β^* as follow:

$$\beta^*(x_j) := \beta_{n+1}(x_j) \qquad \text{for all } j = 0, 1, \ldots, n.$$

The following table gives an idea of how β^* is successively constructed from $\hat{\beta}$. In every new assignment β_i the value of only one variable is changed and this process handles all variables in reverse order. Therefore the rounded value of each variable first appears on the diagonal of the table. This is indicated by printing the corresponding value in bold. This diagonal separates the table into the upper left-hand half with values that are still arbitrary reals indicated by r (the different variables in one row of course generally do not have the same value) and the bottom right-hand half with natural numbers indicated by k. Whereas all values in one row might be different, the column associated with a variable x_i contains at most two different values, and only one if $\hat{\beta}(x_i)$ is already a natural number.

β	$\beta(x_0)$	$\beta(x_1)$	$\cdots$	$\beta(x_{n-i})$	$\beta(x_{n-(i-1)})$	$\cdots$	$\beta(x_{n-1})$	$\beta(x_n)$
$\hat{\beta} := \beta_0$	r	r	$\cdots$	r	r	$\cdots$	r	r
β_1	r	r	$\cdots$	r	r	$\cdots$	r	$\boldsymbol{k}$
β_2	r	r	$\cdots$	r	r	$\cdots$	$\boldsymbol{k}$	k
$\vdots$			$\vdots$			$\vdots$		
β_i	r	r	$\cdots$	r	$\boldsymbol{k}$	$\cdots$	k	k
$\vdots$			$\vdots$			$\vdots$		
β_n	r	$\boldsymbol{k}$	$\cdots$	k	k	$\cdots$	k	k
$\beta^* := \beta_{n+1}$	$\boldsymbol{k}$	k	$\cdots$	k	k	$\cdots$	k	k

Figure 3.5: The successive construction of the assignment β^* from $\hat{\beta}$.

The three assertions about the assignments β_i, $i = 0, \ldots, n+1$ in the next lemma are proved by induction on i. They are then used in turn in the proof of Theorem 3.26.

Lemma 3.28 *For all $i \in \{0, 1, \ldots, n+1\}$ it holds that:*

(a) $\forall c\, (c \in B_\sigma \rightarrow [\![s(c)]\!]_{\beta_i} \in (\, \lfloor [\![s(c)]\!]_{\beta_0} \rfloor - 1 \,,\, \lceil [\![s(c)]\!]_{\beta_0} \rceil + 1 \,) \,)$,

(b) $\forall t\, (t \in T \wedge t^- \leq m_\sigma \rightarrow [\![h_\sigma(t)]\!]_{\beta_i} \leq [\![h_\sigma(t)]\!]_{\beta_0})$,

(c) $\left[\!\!\left[\sum\limits_{k=0}^{n} x_k \right]\!\!\right]_{\beta_i} \leq \left[\!\!\left[\sum\limits_{k=0}^{n} x_k \right]\!\!\right]_{\beta_0}$.

Proof:

Induction on i.

Basis: $i = 0$

For $i = 0$, all three assertions are trivially true.

Step: We assume that the assertions (a), (b) and (c) hold for $1, \ldots, i$, and now prove them for $i + 1$.

If $\beta_i(x_{n-i}) \in \mathbb{N}$, then $\beta_{i+1} = \beta_i$ and thus all assertions follow immediately from the induction hypothesis.

Therefore, we assume that $\beta_i(x_{n-i})$ is not an integer. According to the definition of β_{i+1} there are only two possible values that $\beta_{i+1}(x_{n-i})$ can take:

Case 1: $\beta_{i+1}(x_{n-i}) = \lfloor \beta_i(x_{n-i}) \rfloor$

Hence, it holds that:

$$\beta_{i+1}(x) \leq \beta_i(x) \quad \text{for all} \ \ x \in X_\sigma. \tag{3}$$

For (a): Let b be any inequality in B_σ. If x_{n-i} does not appear in $s(b)$, then $[\![s(b)]\!]_{\beta_{i+1}} = [\![s(b)]\!]_{\beta_i}$, and assertion (a) follows from the induction hypothesis. Hence, assume that x_{n-i} appears in $s(b)$.

Since $\beta_{i+1}(x) \leq \beta_i(x)$ for each $x \in X_\sigma$, it is evident that

$$[\![s(b)]\!]_{\beta_{i+1}} \leq [\![s(b)]\!]_{\beta_i}. \tag{4}$$

By the induction hypothesis, $[\![s(b)]\!]_{\beta_i} < \lceil [\![s(b)]\!]_{\beta_0} \rceil + 1$, so the previous inequality (4) becomes

$$[\![s(b)]\!]_{\beta_{i+1}} < \lceil [\![s(b)]\!]_{\beta_0} \rceil + 1. \tag{5}$$

As $\beta_{i+1}(x_{n-i})$ has been set to $\lfloor \beta_i(x_{n-i}) \rfloor$, the corresponding criterion in the definition of β_{i+1}

$$\forall c (c \in B_\sigma \rightarrow \lfloor [\![s(c)]\!]_{\beta_0} \rfloor - 1 < [\![s(c)]\!]_{\beta_{i+1}})$$

is fulfilled. Since $\underline{\beta_{i+1}} = \beta_{i+1}$, it follows for the inequality b that:

$$\lfloor [\![s(b)]\!]_{\beta_0} \rfloor - 1 < [\![s(b)]\!]_{\beta_{i+1}} \tag{6}$$

and because b has been chosen arbitrarily, the inequalities (5) and (6) together complete the induction step for assertion (a) in this case.

For (b): It follows from the inequality (3) that for each transition t enabled in m_σ

$$[\![h_\sigma(t)]\!]_{\beta_{i+1}} \le [\![h_\sigma(t)]\!]_{\beta_i}$$

and because we know from the induction hypothesis that

$$[\![h_\upsilon(t)]\!]_{\rho_i} \le [\![h_\upsilon(t)]\!]_{\rho_0}$$

assertion (b) is proved in this case.

For (c): The inequality (3) and the induction hypothesis

$\Big[\!\Big[\sum_{k=0}^{n} x_k \Big]\!\Big]_{\beta_i} \le \Big[\!\Big[\sum_{k=0}^{n} x_k \Big]\!\Big]_{\beta_0}$ instantaneously yield $\Big[\!\Big[\sum_{k=0}^{n} x_k \Big]\!\Big]_{\beta_{i+1}} \le \Big[\!\Big[\sum_{k=0}^{n} x_k \Big]\!\Big]_{\beta_0}$,

i.e., *(c)* holds.

Case 2: $\beta_{i+1}(x_{n-i}) = \lceil \beta_i(x_{n-i}) \rceil$

i.e., B_σ contains an inequality $\tilde{c}$ such that

$$[\![s(\tilde{c})]\!]_{\beta_{i+1}} \le \lfloor [\![s(\tilde{c})]\!]_{\beta_0} \rfloor - 1 \tag{7}$$

and x_{n-i} thus appears in $\tilde{c}$.

It also holds in this case that:

$$\beta_i(x) \le \beta_{i+1}(x) \quad \text{for all} \quad x \in X_\sigma. \tag{8}$$

For (a):

Let b be any inequality in B_σ. Then it holds that:

$$\lfloor [\![s(b)]\!]_{\beta_0} \rfloor - 1 \; < \; [\![s(b)]\!]_{\beta_i} \qquad \text{ind. hypothesis}$$
$$\leq \; [\![s(b)]\!]_{\beta_{i+1}} \qquad \text{because of (8)} \qquad (9)$$

On the other hand, it is true for the formula $\tilde{c}$ that:

$$\begin{aligned}
[\![s(\tilde{c})]\!]_{\beta_{i+1}} &= [\![s(\tilde{c})]\!]_{\beta_i} - \beta_i(x_{n-i}) + \beta_{i+1}(x_{n-i}) \\
&= [\![s(\tilde{c})]\!]_{\beta_i} - \beta_i(x_{n-i}) + \lceil \beta_i(x_{n-i}) \rceil \\
&= [\![s(\tilde{c})]\!]_{\beta_i} - \beta_i(x_{n-i}) + \lfloor \beta_i(x_{n-i}) \rfloor + 1 \\
&= [\![s(\tilde{c})]\!]_{\beta_{i+1}} + 1 \\
&\leq \lfloor [\![s(\tilde{c})]\!]_{\beta_0} \rfloor \qquad \text{because of (7)}
\end{aligned}$$

i.e.,

$$[\![s(\tilde{c})]\!]_{\beta_{i+1}} \; \leq \; \lfloor [\![s(\tilde{c})]\!]_{\beta_0} \rfloor \qquad (10)$$
$$\text{and therefore} \qquad [\![s(\tilde{c})]\!]_{\beta_{i+1}} \; \leq \; [\![s(\tilde{c})]\!]_{\beta_0} . \qquad (11)$$

Because of (9) and (10) assertion (a) holds for the formula $\tilde{c}$.

We still need to determine that for every b in B_σ

$$[\![s(b)]\!]_{\beta_{i+1}} \; < \; \lceil [\![s(b)]\!]_{\beta_0} \rceil + 1. \qquad (12)$$

So let us suppose that there is at least one inequality $\tilde{b}$ in B_σ that does not satisfy (12), i.e.,

$$[\![s(\tilde{b})]\!]_{\beta_{i+1}} \geq \lceil [\![s(\tilde{b})]\!]_{\beta_0} \rceil + 1. \qquad (13)$$

This in particular implies

$$[\![s(\tilde{b})]\!]_{\beta_{i+1}} \geq [\![s(\tilde{b})]\!]_{\beta_0} + 1. \qquad (14)$$

Let $j_{\tilde{c}}$ and $k_{\tilde{c}}$ be the minimal and maximal variable indices appearing in $s(\tilde{c})$, respectively. Based on Remark 3.23(iv) it is clear that

$$s(\tilde{c}) = x_{j_{\tilde{c}}} + x_{j_{\tilde{c}}+1} + \ldots + x_{n-i} + x_{n-(i-1)} + \ldots + x_{k_{\tilde{c}}}. \qquad (15)$$

Similarly, let $j_{\tilde{b}}$ and $k_{\tilde{b}}$ be the minimal and maximal variable indices appearing in $s(\tilde{b})$, respectively. Then $s(\tilde{b})$ has the form:

$$s(\tilde{b}) = x_{j_{\tilde{b}}} + x_{j_{\tilde{b}}+1} + \ldots + x_{n-i} + x_{n-(i-1)} + \ldots + x_{k_{\tilde{b}}}. \qquad (16)$$

Hence, it holds for the indices $n - i, k_{\tilde{c}}$ and k_b that:

$$n - i \leq k_{\tilde{c}} \qquad \text{and} \qquad n - i \leq k_{\tilde{b}},$$
$$\text{i.e.,} \quad n - k_{\tilde{c}} < i + 1 \qquad \text{and} \qquad n - k_{\tilde{b}} < i + 1. \qquad (17)$$

The kind of values of the variables - reals or integers - under the assignments β_0 and β_{i+1} are:

$$\underbrace{\beta_{i+1}(x_{j_{\tilde{c}}}) \quad \cdots \quad \beta_{i+1}(x_{n-(i+1)})}_{\substack{\beta_{i+1} \ = \ \beta_0 \\ \uparrow \qquad \uparrow \\ \text{real} \qquad \text{real}}} \underbrace{\beta_{i+1}(x_{n-i}) \quad \cdots \quad \beta_{i+1}(x_{k_{\tilde{c}}})}_{\substack{\beta_{i+1} \ \neq \ \beta_0 \\ \uparrow \qquad \uparrow \\ \text{integer} \qquad \text{real}}} \qquad (18)$$

According to the definition of β^* the assignment β_{i+1} changes the value of the variable x_{n-i} and for every r with $-1 \leq r \leq n-1$ the assignment β_{n-r} changes the value of x_{r+1}.

Hence, according to (15), we may rewrite (11) as

$$\begin{aligned}
&(\beta_{i+1}(x_{j_{\tilde{c}}}) - \beta_0(x_{j_{\tilde{c}}})) + \\
&(\beta_{i+1}(x_{j_{\tilde{c}}+1}) - \beta_0(x_{j_{\tilde{c}}+1})) + \ldots + \\
&(\beta_{i+1}(x_{n-i}) - \beta_0(x_{n-i})) + (\beta_{i+1}(x_{n-i+1}) - \beta_0(x_{n-i+1})) + \ldots + \\
&(\beta_{i+1}(x_{k_{\tilde{c}}}) - \beta_0(x_{k_{\tilde{c}}})) \leq 0.
\end{aligned} \qquad (19)$$

and according to (18), this may be rewritten as

$$\begin{aligned}
&(\beta_{i+1}(x_{n-i}) - \beta_0(x_{n-i})) + (\beta_{i+1}(x_{n-i+1}) - \beta_0(x_{n-i+1})) + \ldots + \\
&(\beta_{i+1}(x_{k_{\tilde{c}}}) - \beta_0(x_{k_{\tilde{c}}})) \leq 0.
\end{aligned} \qquad (20)$$

Similarly, (14) and (16) yield

$$(\beta_{i+1}(x_{n-i}) - \beta_0(x_{n-i})) + (\beta_{i+1}(x_{n-i+1}) - \beta_0(x_{n-i+1})) + \ldots +$$
$$(\beta_{i+1}(x_{k_{\tilde{b}}}) - \beta_0(x_{k_{\tilde{b}}})) \geq 1. \tag{21}$$

We now consider the relationship between the two maximal indices $k_{\tilde{c}}$ and $k_{\tilde{b}}$. There are three possibilities:

Case 2.1: $k_{\tilde{c}} = k_{\tilde{b}}$
    Then

$$
\begin{aligned}
1 &\leq [\![s(\tilde{b})]\!]_{\beta_{i+1}} - [\![s(\tilde{b})]\!]_{\beta_0} && \text{because of (14)} \\
&= [\![s(\tilde{c})]\!]_{\beta_{i+1}} - [\![s(\tilde{c})]\!]_{\beta_0} && \text{because of (20) and (21)} \\
&\leq 0 && \text{because of (11)}
\end{aligned}
$$

.   which is clearly a contradiction.

Case 2.2: $k_{\tilde{c}} < k_{\tilde{b}}$
    In this case the two terms $s(\tilde{c})$ and $s(\tilde{b})$ have the form:

$$
\begin{aligned}
s(\tilde{c}) &= x_{j_{\tilde{c}}} + \cdots \quad + x_{n-i} + \quad \cdots + x_{k_{\tilde{c}}} \\
s(\tilde{b}) &= x_{j_{\tilde{b}}} + \cdots \quad + x_{n-i} + \quad \cdots + x_{k_{\tilde{c}}} + \cdots + x_{k_{\tilde{b}}}.
\end{aligned}
$$

We now consider the type of values of the variables appearing in $s(\tilde{b})$ under the assignments $\beta_0, \beta_{n-k_{\tilde{c}}}$ and β_{i+1}. Because of (1) and (2) it holds that:

$s(\tilde{b})\;=$

$$
\underbrace{x_{j_{\tilde{b}}} + \cdots + x_{n-i-1}}_{\substack{\beta_{i+1} \;=\; \beta_{n-k_{\tilde{c}}} \\ \uparrow \qquad\quad \uparrow \\ \text{real} \qquad \text{real}}}\;\;
\overbrace{\underbrace{x_{n-i} + \cdots + x_{k_{\tilde{c}}}}_{\substack{\beta_{i+1} \;\neq\; \beta_{n-k_{\tilde{c}}} \\ \uparrow \qquad\quad \uparrow \\ \text{integer} \qquad \text{real}}}}^{\beta_0 = \beta_{n-k_{\tilde{c}}}}\;\;
\overbrace{\underbrace{x_{k_{\tilde{c}}+1} + \cdots + x_{k_{\tilde{b}}}}_{\substack{\beta_{i+1} \;=\; \beta_{n-k_{\tilde{c}}} \\ \uparrow \qquad\quad \uparrow \\ \text{integer} \qquad \text{integer}}}}^{\beta_0 \neq \beta_{n-k_{\tilde{c}}}}\,. \tag{22}
$$

Because of (22), β_0 and $\beta_{n-k_{\tilde{c}}}$ agree on all variables with indices at most $k_{\tilde{c}}$. Hence, the inequality (20) leads to

$$(\beta_{i+1}(x_{n-i}) - \beta_{n-k_{\tilde{c}}}(x_{n-i}))+$$
$$(\beta_{i+1}(x_{n-i+1}) - \beta_{n-k_{\tilde{c}}}(x_{n-i+1})) + \ldots +$$
$$(\beta_{i+1}(x_{k_{\tilde{c}}}) - \beta_{n-k_{\tilde{c}}}(x_{k_{\tilde{c}}})) \leq 0. \tag{23}$$

Thus, (22) and (23) yield

$$[\![s(\tilde{b})]\!]_{\beta_{i+1}} - [\![s(\tilde{b})]\!]_{\beta_{n-k_{\tilde{c}}}} \leq 0. \tag{24}$$

But (13) and (24) then yield

$$[\![s(\tilde{b})]\!]_{\beta_{n-k_{\tilde{c}}}} \geq \lceil [\![s(\tilde{b})]\!]_{\beta_0} \rceil + 1$$

which contradicts the induction hypothesis for $n - k_{\tilde{c}}$, because $n - k_{\tilde{c}} < i + 1$ (cf. (17)).

Case 2.3: $k_{\tilde{c}} > k_{\tilde{b}}$
Then, $s(\tilde{c})$ and $s(\tilde{b})$ have the form:

$$s(\tilde{c}) = x_{j_{\tilde{c}}} + \cdots + x_{n-i} + \cdots + x_{k_{\tilde{b}}} + \cdots x_{k_{\tilde{c}}}$$
$$s(\tilde{b}) = x_{j_{\tilde{b}}} + \cdots + x_{n-i} + \cdots + x_{k_{\tilde{b}}}$$

Analogously to Case 2.2 this leads to the following type of values of the variables in $s(\tilde{c})$ under the assignments $\beta_0, \beta_{n-k_{\tilde{b}}}$ and β_{n+1}:

$$s(\tilde{c}) =$$

$\overbrace{x_{j_{\tilde{b}}} + \cdots + x_{n-i-1}}^{\beta_0=\beta_{n-k_{\tilde{b}}}}$	$x_{n-i} + \cdots + x_{k_{\tilde{b}}}$	$\overbrace{x_{k_{\tilde{b}}+1} + \cdots + x_{k_{\tilde{c}}}}^{\beta_0 \neq \beta_{n-k_{\tilde{b}}}}$	$.(25)$

$$\begin{array}{cccccc} \beta_{i+1} & = & \beta_{n-k_{\tilde{b}}} & \beta_{i+1} & \neq & \beta_{n-k_{\tilde{b}}} & \beta_{i+1} & = & \beta_{n-k_{\tilde{b}}} \\ \uparrow & & \uparrow & \uparrow & & \uparrow & \uparrow & & \uparrow \\ \text{real} & & \text{real} & \text{integer} & & \text{real} & \text{integer} & & \text{integer} \end{array}$$

According to (25), β_0 and $\beta_{n-k_{\tilde{b}}}$ agree on all variables with indices at most $k_{\tilde{b}}$. Therefore inequality (21) leads to

$$(\beta_{i+1}(x_{n-i}) - \beta_{n-k_{\tilde{b}}}(x_{n-i})) +$$
$$(\beta_{i+1}(x_{n-i+1}) - \beta_{n-k_{\tilde{b}}}(x_{n-i+1})) + \ldots + \tag{26}$$
$$(\beta_{i+1}(x_{k_{\tilde{b}}}) - \beta_{n-k_{\tilde{b}}}(x_{k_{\tilde{b}}})) \geq 1.$$

Hence, (26) together with (15) shows that

$$[\![s(\tilde{c})]\!]_{\beta_{i+1}} - [\![s(\tilde{c})]\!]_{\beta_{n-k_{\tilde{b}}}} \geq 1. \tag{27}$$

But (10) and (27) then yield

$$[\![s(\tilde{c})]\!]_{\beta_{n-k_{\tilde{b}}}} \leq \lfloor [\![s(\tilde{c})]\!]_{\beta_0} \rfloor - 1$$

which contradicts the induction hypothesis for $n - k_{\tilde{b}}$, because $n - k_{\tilde{b}} < i + 1$ (cf. (17)).

So assumption (13) has led to contradictions in all three sub-cases. Therefore the following inequality must hold for all b in B_σ:

$$[\![s(b)]\!]_{\beta_{i+1}} < \lceil [\![s(b)]\!]_{\beta_0} \rceil + 1 \tag{28}$$

Thus, (9) and (28) prove assertion (a) in this case.

For (b):

Let t be a transition which is enabled after the firing of σ. If x_{n-i} does not appear in $h_\sigma(t)$, then $[\![h_\sigma(t)]\!]_{\beta_{i+1}} = [\![h_\sigma(t)]\!]_{\beta_i}$, and $[\![h_\sigma(t)]\!]_{\beta_{i+1}} \leq [\![h_\sigma(t)]\!]_{\beta_0}$ follows from the induction hypothesis. Therefore, assume that x_{n-i} does appear in $h_\sigma(t)$.

According to Remark 3.23.(i) (applied to h_σ), the variable x_n appears in every component of h_σ which is not $\sharp$. Together with Remark 3.23.(ii), this implies that there is an index j_t such that

$$h_\sigma(t) = x_{j_t} + x_{j_t+1} + \ldots + x_{n-i} + x_{n-(i-1)} + \ldots + x_n. \tag{29}$$

Because of (22), β_{i+1} and $\beta_{n-k_{\tilde{c}}}$ agree on all variables with indices smaller than $n - i$ and on all variables with indices greater than $k_{\tilde{c}}$.

Hence, (23) together with (29) shows that

$$[\![h_\sigma(t)]\!]_{\beta_{i+1}} - [\![h_\sigma(t)]\!]_{\beta_{n-k_{\tilde{c}}}} \leq 0. \tag{30}$$

Using the induction hypothesis for $n - k_{\tilde{c}}$, (30) yields

$$[\![h_\sigma(t)]\!]_{\beta_{i+1}} \leq [\![h_\sigma(t)]\!]_{\beta_0}.$$

Since t was chosen arbitrarily, the assertion (b) is also proved in this case.

For (c):

Again, because of (22), β_{i+1} and $\beta_{n-k_{\tilde{c}}}$ agree on all variables with indices smaller than $n - i$, and on all variables with indices greater than $k_{\tilde{c}}$ it follows from (23) that

$$[\![\sum_{k=0}^{n} x_k]\!]_{\beta_{i+1}} \leq [\![\sum_{k=0}^{n} x_k]\!]_{\beta_{n-k_{\tilde{c}}}}. \tag{31}$$

And by the induction hypothesis for $n - k_{\tilde{c}}$

$$[\![\sum_{k=0}^{n} x_k]\!]_{\beta_{n-k_{\tilde{c}}}} \leq [\![\sum_{k=0}^{n} x_k]\!]_{\beta_0}$$

which, together with the inequality (31), proves the third assertion (c):

$$[\![\sum_{k=0}^{n} x_k]\!]_{\beta_{i+1}} \leq [\![\sum_{k=0}^{n} x_k]\!]_{\beta_0}.$$

With this Lemma 3.28 is proved. □

Proof of Theorem 3.26:

With $\beta^* := \beta_{n+1}$ assertion (2) of Theorem 3.26 is the same as (b) of Lemma 3.28 and (3) of Theorem 3.26 is the same as (c) of Lemma 3.28.

So in order to prove item *(1)* of Theorem 3.26 we now show that the values $\beta^*(x_0), \ldots \ldots, \beta^*(x_n)$ satisfy all inequalities in B_σ.

Thus, let c be an arbitrary inequality in B_σ. Since $\beta^*(x_i)$, $i = 0, \ldots, n$ are integers it is clear that $[\![s(c)]\!]_{\beta^*}$ is also an integer. Assertion (a) from Lemma 3.28 implies that

$$[\![s(c)]\!]_{\beta^*} \in (\ \lfloor [\![s(c)]\!]_{\beta_0} \rfloor - 1, \lceil [\![s(c)]\!]_{\beta_0} \rceil + 1\ ).$$

But the only integers in the interval $(\lfloor [\![s(c)]\!]_{\hat{\beta}_0}\rfloor - 1, \lceil [\![s(c)]\!]_{\hat{\beta}_0}\rceil + 1)$ are $\lfloor [\![s(c)]\!]_{\hat{\beta}_0}\rfloor$ and $\lceil [\![s(c)]\!]_{\hat{\beta}_0}\rceil$. This means that

$$[\![s(c)]\!]_{\beta^*} = \lfloor [\![s(c)]\!]_{\hat{\beta}_0}\rfloor \text{ or } [\![s(c)]\!]_{\beta^*} = \lceil [\![s(c)]\!]_{\hat{\beta}_0}\rceil. \tag{32}$$

The inequality c has the form $s(c) \leq k$ or $k \leq s(c)$ for some integer k (more precisely: According to the definition of B_σ, k is 0 or $eft(t)$ or $lft(t)$ for some transition t in the considered Time Petri net).

We now first consider the case that c has the form $s(c) \leq k$. Since the first assignment $\hat{\beta}$ defines a feasible run, all inequalities in B_σ are satisfied by the values $\hat{\beta}(x_i)$, x_i for all $i = 0, \cdots, n$. It follows that

$$[\![s(c)]\!]_{\hat{\beta}_0} \leq k,$$

which implies

$$\begin{aligned} \lceil [\![s(c)]\!]_{\hat{\beta}_0}\rceil &\leq \lceil k \rceil \\ &= k. \end{aligned}$$

Thus, (32) leads to

$$[\![s(c)]\!]_{\beta^*} \leq k,$$

which means that the natural numbers $\beta^*(x_0), \ldots, \beta^*(x_n)$ fulfill the inequality c.

If c has the form $k \leq s(c)$ we can similarly prove that the natural numbers $\beta^*(x_0), \ldots, \beta^*(x_n)$ fulfill c.

Since c was chosen arbitrarily, β^* satisfies all conditions in B_σ, so the statement (1) and hence Theorem 3.26 is proved.   □

In the following example we demonstrate how a feasible run with only integer times elapsing between firings is constructed from an arbitrary feasible run.

Example 3.29 *Consider the Time Petri net* $\mathcal{Z}_3$

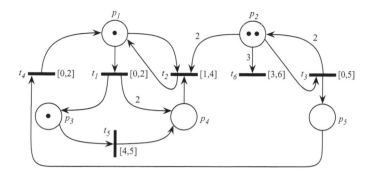

Figure 3.6: The Time Petri net $\mathcal{Z}_3$

and the transition sequence $\sigma = t_1 t_3 t_4 t_2 t_3$. *The run*

$$\sigma(\tau) := z_0 \xrightarrow{0.7} \xrightarrow{t_1} \xrightarrow{0.0} \xrightarrow{t_3} \xrightarrow{0.4} \xrightarrow{t_4} \xrightarrow{1.2} \xrightarrow{t_2} \xrightarrow{0.5} \xrightarrow{t_3} \xrightarrow{1.4} z$$

is feasible in $\mathcal{Z}_3$. *The elapsed times are* $\tau = (\ 0.7,\ 0.0,\ 0.4,\ 1.2,\ 0.5,\ 1.4\ )$.

For the transition sequence σ *the parametric run* $\sigma(x)$ *has the form* $\sigma(x) = (x_0, t_1, x_1, t_3, x_2, t_4, x_3, t_2, x_4, t_3, x_5)$ *and we obtain the parametric state* (z_σ, B_σ) *with* $z_\sigma = (m_\sigma, h_\sigma)$ *with*

$$m_\sigma = (1, 2, 2, 1, 1),$$

$$h_\sigma = \begin{pmatrix} x_4 + x_5 \\ x_5 \\ x_5 \\ x_5 \\ x_0 + x_1 + x_2 + x_3 + x_4 + x_5 \\ \natural \end{pmatrix} \quad and$$

$$B_\sigma = \left\{ \begin{array}{lll} 0 \le x_0, & x_0 \le 2, & x_0 + x_1 + x_2 \le 5, \\ 0 \le x_1, & x_2 \le 2, & x_2 + x_3 \le 5, \\ 0 \le x_2, & x_3 \le 2, & x_0 + x_1 + x_2 + x_3 \le 5, \\ 1 \le x_3, & x_4 \le 2, & x_0 + x_1 + x_2 + x_3 + x_4 \le 5, \\ 0 \le x_4, & x_5 \le 2, & x_0 + x_1 + x_2 + x_3 + x_4 + x_5 \le 5, \\ 0 \le x_5, & x_0 + x_1 \le 5, & x_4 + x_5 \le 2 \end{array} \right\}.$$

Omitting all redundant inequalities from B_σ (18 inequalities) we obtain a smaller system of inequalities (11 inequalities) with the same solution set. Thus, instead of B_σ we consider the following system without redundant inequalities:

$$\left\{ \begin{array}{ll} 0 \le x_0, & x_0 \le 2, \\ 0 \le x_1, & x_2 \le 2, \\ 0 \le x_2, & x_3 \le 2, \\ 1 \le x_3, & x_0 + x_1 + x_2 + x_3 + x_4 + x_5 \le 5, \\ 0 \le x_4, & x_4 + x_5 \le 2, \\ 0 \le x_5 \end{array} \right\}.$$

We obtain the first assignment $\hat{\beta}$ for the variables $x_0, x_1, x_2, x_3, x_4, x_5$ from τ, i.e., $\hat{\beta}(x_0) = 0.7$, $\hat{\beta}(x_1) = 0.0$, $\hat{\beta}(x_2) = 0.4$, $\hat{\beta}(x_3) = 1.2$, $\hat{\beta}(x_4) = 0.5$, $\hat{\beta}(x_5) = 1.4$. In seven steps we now recursively compute the values $\beta^*(x_i) = \beta_6(x_i)$ from the values $\beta_0(x_i) := \hat{\beta}(x_i)$. Here it depends on the intervals $(\lfloor [\![s(c)]\!]_{\beta_0} \rfloor - 1, \lceil [\![s(c)]\!]_{\beta_0} \rceil + 1)$ for all $c \in B_\sigma$[4] whether the value of a variable is rounded up or down.

For instance, for the inequality $c = x_0 + x_1 + x_2 + x_3 + x_4 + x_5 \le 5$ the interval $(\lfloor [\![s(c)]\!]_{\beta_0} \rfloor - 1, \lceil [\![s(c)]\!]_{\beta_0} \rceil + 1)$ is $(3, 6)$, for $c = x_4 + x_5 \le 2$ the interval is $(0, 3)$, and for $c = x_0 \le 2$ the interval is $(-1, 2)$ which is also the interval for the inequality $c = 0 \le x_0$.

Note that the assignments β_i for $1 < i < n + 1$ might not define solutions for the whole set of inequalities B_σ. Only $\hat{\beta}$ and β^* are guaranteed to define solutions for B_σ.

---

[4]As mentioned above, it is sufficient to consider only the system of non-redundant inequalities in B_σ.

*In the next table I the recursive construction of integer values for $x_0, x_1, x_2, x_3,$
x_4, x_5 according to Theorem 3.26 is shown. Since $h_\sigma(t_2) = h_\sigma(t_3) = h_\sigma(t_4)$
only $h_\sigma(t_2)$ of the three is shown in the table.*

I		x_0	x_1	x_2	x_3	x_4	x_5	$h_\sigma(t_1)$	$h_\sigma(t_2)$	$h_\sigma(t_5)$
$\hat\beta$ =	β_0	0.7	0.0	0.4	1.2	0.5	1.4	1.9	1.4	4.2
	β_1	0.7	0.0	0.4	1.2	0.5	**1**	1.5	1.0	3.8
	β_2	0.7	0.0	0.4	1.2	**0**	1	1.0		3.3
	β_3	0.7	0.0	0.4	**1**	0	1			3.1
	β_4	0.7	0.0	**1**	1	0	1			3.7
	β_5	0.7	**0**	1	1	0	1			3.7
β^* =	β_6	**1**	0	1	1	0	1			4.0

Thus, we obtain the feasible run $\sigma(\tau_1^*)$ in $\mathcal{Z}_3$ with

$$\sigma(\tau_1^*) := \underline{z_0} \xrightarrow{1} \underline{} \xrightarrow{t_1} \xrightarrow{0} \underline{} \xrightarrow{t_3} \xrightarrow{1} \underline{} \xrightarrow{t_4} \xrightarrow{1} \underline{} \xrightarrow{t_2} \xrightarrow{0} \underline{} \xrightarrow{t_3} \xrightarrow{1} \underline{z}.$$

Furthermore, it holds that $\underline{z} = (\underline{m}, \underline{h}) := (m_\sigma, \beta^*(h_\sigma))$, i.e., $\underline{h} = (1, 1, 1, 1, 4, \sharp)$.

We now state **one of the most important properties** of Time Petri
nets which follows from Theorem 3.26:

Remark 3.30 *Let m^* be an arbitrary reachable p marking in a Time Petri
net $\mathcal{Z}$. Then, m^* is reachable in $\mathcal{Z}$ with a feasible integer-run. The state
$z = (m, h)$ with $m = m^*$, reached after firing this run, is an integer-state.*

Corollary 3.31 *Let $z = (m, h)$ be an arbitrary reachable state in a Time
Petri net $\mathcal{Z}$. Then the state $\underline{z} := (m, \lfloor h \rfloor)$ is also reachable in $\mathcal{Z}$.*

Proof: It follows immediately from Theorem 3.26 together with Lemma
3.28.(a) that $\underline{z} := (m, \lfloor h \rfloor)$ is a reachable state in $\mathcal{Z}$. $\square$

The next theorem is a counterpart to Theorem 3.26:

Theorem 3.32 *Let $\mathcal{Z} = (P, T, F, V, m_0, I)$ be a Time Petri net, σ a transition sequence of length n with $z_0 \xrightarrow{\sigma} (z_\sigma, B_\sigma)$, $z_\sigma = (m_\sigma, h_\sigma)$, and $X_\sigma = \{x_0, x_1, \ldots, x_n\}$. Furthermore let $\sigma(\hat{\beta}(x))$ with $\hat{\beta} : X_\sigma \longrightarrow \mathbb{R}_0^+$ be a feasible run of σ. Then, there exists an assignment $\beta^* : X_\sigma \longrightarrow \mathbb{N}$ for the variables such that:*

(1) $\sigma(\beta^*(x))$ is also a feasible run of σ in $\mathcal{Z}$,

(2) for each transition t enabled in m_σ it holds that $[\![h_\sigma(t)]\!]_{\beta^*} \geq [\![h_\sigma(t)]\!]_{\hat{\beta}}$,

(3) the total duration of the run $\sigma(\beta^*(x))$ is no smaller than the total duration of the run $\sigma(\hat{\beta}(x))$, i.e., $\left[\!\!\left[\sum\limits_{k=0}^{n} x_k \right]\!\!\right]_{\beta^*} \geq \left[\!\!\left[\sum\limits_{k=0}^{n} x_k \right]\!\!\right]_{\hat{\beta}}$.

The construction of the assignment β^* is "dual" to the construction of β^* for Theorem 3.26[5]: We define the assignment β_1 by rounding up and continue to round up unless the value of an inequality c in B_σ would thereby become larger than $\lceil [\![s(c)]\!]_{\beta_0} \rceil + 1$, in which case we round down. To prove Theorem 3.32 we would state a lemma which is a counterpart to Lemma 3.28. However we abstain from a proof here because it is similar to the proof of Theorem 3.26.

Example 3.33 *Let us again consider the Time Petri net $\mathcal{Z}_3$, the transition sequence σ and the feasible run $\sigma(\tau)$ from Example 3.29.*

In the following table the recursive construction of integer values for $x_0, x_1, x_2, x_3, x_4, x_5$ in $\sigma(x) = x_0 t_1 x_1 t_3 x_2 t_4 x_3 t_2 x_4 t_3 x_5$ according to Theorem 3.32 is shown.

II		x_0	x_1	x_2	x_3	x_4	x_5	$h_\sigma(t_1)$	$h_\sigma(t_2)$	$h_\sigma(t_5)$
$\hat{\beta} =$	β_0	0.7	0.0	0.4	1.2	0.5	1.4	1.9	1.4	4.2
	β_1	0.7	0.0	0.4	1.2	0.5	**2**	2.5	2.0	4.8
	β_2	0.7	0.0	0.4	1.2	**0**	2	2.0		4.3
	β_3	0.7	0.0	0.4	**2**	0	2			5.1
	β_4	0.7	0.0	**1**	2	0	2			5.7
	β_5	0.7	**0**	1	2	0	2			5.7
$\beta^* =$	β_6	**0**	0	1	2	0	2			5.0

[5]cf. Exercise 3.5

We obtain the feasible run $\sigma(\tau_2^*)$ in $\mathcal{Z}_3$:

$$\sigma(\tau_2^*) := z_0 \xrightarrow{0} \xrightarrow{t_1} \xrightarrow{0} \xrightarrow{t_3} \xrightarrow{1} \xrightarrow{t_4} \xrightarrow{2} \xrightarrow{t_2} \xrightarrow{0} \xrightarrow{t_3} \xrightarrow{2} \overline{z}$$

for the integer-state $\overline{z} = (\overline{m}, \overline{h}) = (m_\sigma, \beta^*(h))$ with $\overline{h} = (2, 2, 2, 2, 5, \natural)$.

The next corollary now follows immediately:

Corollary 3.34 *Let $z = (m, h)$ be an arbitrary reachable state in a Time Petri net $\mathcal{Z}$. Then the state $\overline{z} := (m, \lceil h \rceil)$ is also reachable in $\mathcal{Z}$.*

Combining Corollaries 3.31 and 3.34 we obtain a necessary condition for the reachability of a state in a Time Petri net:

Remark 3.35 (A necessary reachability condition)
Let z be an arbitrary reachable state in a Time Petri net $\mathcal{Z}$. Then, the states $\underline{z}$ and $\overline{z}$ are also reachable in $\mathcal{Z}$.

In other words, if at least one of the integer-states $\underline{z}$ or $\overline{z}$ is not reachable in $\mathcal{Z}$ then z is not reachable in $\mathcal{Z}$ either.

But, as we will see in the next example, this condition is not sufficient.

Example 3.36 *Let us consider the Time Petri net $\mathcal{Z}_4$ in Fig. 3.7 and the state $z^* = (m^*, h^*)$ with $m^* = (0, 1, 1, 0, 1, 0)$ and $h^* = (\natural, 1.2, 1.5, \natural, \natural, 0.3)$. The states $\lfloor z^* \rfloor = (m^*, (\natural, 1, 1, \natural, \natural, 0))$ and $\lceil z^* \rceil = (m^*, (\natural, 2, 2, \natural, \natural, 1))$ are respectively reachable in $\mathcal{Z}_4$ through*

$$z_0 \xrightarrow{0\,t_1\,1\,t_4} \lfloor z^* \rfloor$$

and

$$z_0 \xrightarrow{0\,t_1\,1\,t_4\,1} \lceil z^* \rceil.$$

The state z^* on the other hand is not reachable in $\mathcal{Z}$, because the transitions t_2 and t_3 are always enabled together and therefore the time since their last enabling is always the same, i.e., as long as they are both enabled their clocks

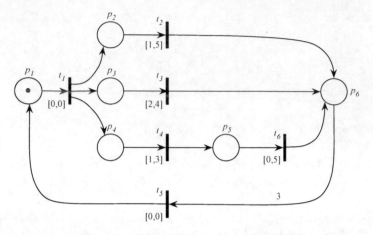

Figure 3.7:   The Time Petri net $\mathcal{Z}_4$

always show the same time.

We give another easy example with the next small Time Petri net $\mathcal{Z}_{4A}$ in Fig. 3.8.

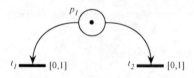

Figure 3.8:   The Time Petri net $\mathcal{Z}_{4A}$

It is clear that the integer-states $z_1 = \big((1),(0,0)\big)$ and $z_2 = \big((1),(1,1)\big)$ are reachable in $\mathcal{Z}_{4A}$ and the state $z^* = \big((1),(0.2,0.3)\big)$ is not reachable in $\mathcal{Z}_{4A}$. Note that $\lfloor z^* \rfloor = z_1$ and $\lceil z^* \rceil = z_2$.

We can now easily prove the next property which together with Theorems 3.26 and 3.32 is of fundamental importance for the analysis of Time Petri nets.

In the following we will call a Time Petri net *finite*, if for every transition t in the net $lft(t)$ is a natural number (and not ∞).

Theorem 3.37 *Let $\mathcal{Z}$ be a finite Time Petri net. Then the set $RIS_{\mathcal{Z}}$ of all reachable integer-states in $\mathcal{Z}$ is finite if and only if the set $R_{\mathcal{Z}}$ of all reachable p-markings in $\mathcal{Z}$ is finite.*

Proof:

$(\Longrightarrow)$

Let the set $RIS_{\mathcal{Z}}$ be finite. According to Theorem 3.26 there exists a reachable integer-state $\tilde{z} = (\tilde{m}, h) \in RIS_{\mathcal{Z}}$ for each reachable p-marking $\tilde{m}$ in $\mathcal{Z}$. It is clear that for two different p-markings $\tilde{m}_1$ and $\tilde{m}_2$ there are also two different integer-states $\tilde{z}_1$ und $\tilde{z}_2$, i.e., the set $RIS_{\mathcal{Z}}$ has at least as many elements as $R_{\mathcal{Z}}$. Because we assumed that $RIS_{\mathcal{Z}}$ is finite it follows immediately that $R_{\mathcal{Z}}$ is finite, too.

$(\Longleftarrow)$

Let the set $R_{\mathcal{Z}}$ of all reachable p-markings in $\mathcal{Z}$ be finite. Let m' be an arbitrary p-marking in $R_{\mathcal{Z}}$. Because $\mathcal{Z}$ is finite, the set

$$H_{m'} := \{\, (h(t)) \in (\mathbb{N} \cup \{\natural\})^{|T|} \mid$$
$$(h(t) \in ([0, lft(t)] \cap \mathbb{N}) \,,\, \text{if } t^- \leq m') \wedge$$
$$(h(t) = \natural \,,\, \text{if } t^- \not\leq m') \,\}$$

of all integer-t-markings in $\mathcal{Z}$ suitable for m' is finite, too, because for each transition $t \in T$ the set $[0, lft(t)] \cap \mathbb{N}$ is finite. Hence, the set

$$HR_{m'} := \{\, (h(t)) \in (\mathbb{N} \cup \{\natural\})^{|T|} \mid \exists z\,(\, z = (m', h) \,\wedge\, z \text{ reachable in } \mathcal{Z})\}$$

which is a subset of $H_{m'}$ is finite as well. Therefore the set

$$Z_{m'} := \{z = (m', h') \mid h' \in HR_{m'}\}$$

is also finite. Furthermore, because of Theorem 3.26 it holds that:

$$RIS_{\mathcal{Z}} = \bigcup_{m \in R_{\mathcal{Z}}} Z_m.$$

Thus, $RIS_{\mathcal{Z}}$ is a finite union of finite sets and therefore is finite itself. $\qquad\square$

3.5   Reachability Graphs for Finite Time Petri Nets

In general, the state space of a Time Petri net can be infinite. Every state is a pair of two tuples, the p-marking with only natural numbers as components and the t-marking with its arbitrary non-negative real number components. In the previous section we have proved that in order to decide the reachability of a p-marking it is sufficient to take into account only integer-states. This does not mean that reachability in Time Petri nets is decidable. But it tells us that the integer-states contain sufficient information for studying this problem which is central in the analysis of any kind of Petri net. In this section we define reachability graphs of finite Time Petri nets. As the integer-states carry all necessary information about reachability we will define the vertices of such a graph to be the integer-states of the Petri net and there will be a directed edge from such a vertex z_1 to the vertex z_2 if the state z_1 can change into z_2 by firing a single transition or by elapsing of time in the net. The edge from z_1 to z_2 is labeled with the transition being fired or the natural number denoting the amount of elapsed time, respectively.

It is clear that such a graph is infinite for any net where at least one transition t has no latest firing time, i.e., $lft(t) = \infty$. However, assuming the latest firing time of every transition in the considered net is a natural number we can conclude with Proposition 3.37 that the graph is finite if and only if the set of all p-markings reachable in the considered net is finite. We will consistently generalize the notion *reachability graph* to arbitrary Time Petri nets in the next section.

We will furthermore see that the integer-states provide sufficient information for studying not only reachability in a Time Petri net but also liveness, reversibility, and more or less all qualitative properties. They also enable us to answer some quantitative questions, for example about the earliest or latest possible time for reaching a certain p-marking or state. We will discuss such quantitative problems later in this chapter.

In the following definition we recursively define the *reachability graph of a Time Petri net*.

Definition 3.38 (reachability graph) *Let $\mathcal{Z} = (P, T, F, V, m_0, I)$ be a finite Time Petri net. The reachability graph of $\mathcal{Z}$ $\mathcal{RG}_{\mathcal{Z}} := (W, E, T \cup \{1\})$*

is the directed graph with edge labels with set of vertices W, set of edges E and edge labels from $T \cup \{1\}$ defined by the following algorithm:

Basis: $W := \{z_0\}$, $E := \emptyset$.

Step: *For each $z \in W$:*

   (1) For each $t \in T$ such that t is ready to fire in z compute the z' such that $z \xrightarrow{t} z'$;
       $W := W \cup \{z'\}$;
       $E := E \cup \{(z, t, z')\}$.

   (2) If $z \xrightarrow{1}$ is feasible in $\mathcal{Z}$, then compute the z' such that $z \xrightarrow{1} z'$;
       $W := W \cup \{z'\}$;
       $E := E \cup \{(z, 1, z')\}$.

Referring to Theorem 3.26 and Remark 3.30 it is easily verified that:

$$W = RIS_{\mathcal{Z}}.$$

In the following we want to reduce the reachability graph defined above. The idea is to "fuse together" consecutive edges representing the elapse of single time units with the transition firing after these elapses into one edge labeled with the transition and the sum of these elapses.

Let us consider the (part of a) run

$$\xrightarrow{t_1} z_i \xrightarrow{1} z_{i+1} \xrightarrow{1} \cdots \xrightarrow{1} z_{i+n} \xrightarrow{t_2} z_{i+(n+1)},$$

and the path in the reachability graph representing it, cf. Fig. 3.9. We are interested in paths beginning and ending with edges labeled with transitions of which all internal edges are labeled with 1. The vertices on the path might of course also have other input and output edges in addition to the ones on the path.

We will replace such a path by a shorter path similar to the one corresponding to the following run:

$$\xrightarrow{t_1} z_i \xrightarrow{n,t_2} z_{i+(n+1)}.$$

Note that the vertices z_{i+j} for $j \in \{1, ..., n\}$ are omitted only if they become isolated[6] when all edges of the path except for the input edge of z_i are ignored.

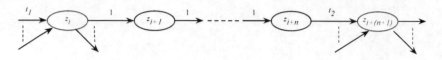

Figure 3.9: Part of a path in the (unreduced) reachability graph

We "fuse" edges and vertices along the path into a new edge with a new label and consider this to be a reduction because the reduced graph generally contains fewer vertices than the original one and often also considerably fewer edges. There are nonetheless cases where the reduction increases the number of edges.

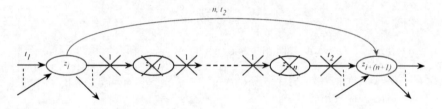

Figure 3.10: Idea for the reduction of the reachability graph of a Time Petri net: Replacing the path from Fig. 3.9

In the above example we would for instance introduce an edge from z_i to $z_{i+(n+1)}$ and label it with n, t_2.

If however a state z_{i+k} (for some k with $1 \leq k \leq n$) has another predecessor in the graph $\mathcal{RG}_{\mathcal{Z}}$, apart from z_{i+k-1}, or another successor state z_r, apart from z_{i+k+1}, their edges are then labeled by transitions (cf. Fig. 3.11) and

---

[6]A vertex is *isolated* if it has neither input nor output edges.

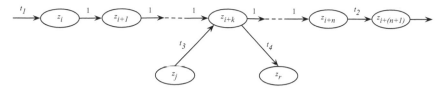

Figure 3.11: Part of a path in the (unreduced) reachability graph with branching vertex

we need to reduce the paths from z_{i+k} to $z_{i+(n+1)}$ and from z_i to z_r (via z_{i+k}) respectively (cf. Fig. 3.12).

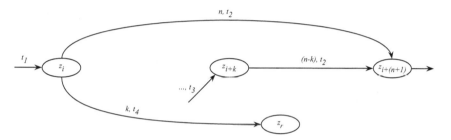

Figure 3.12: The idea for reduction of the reachability graph of a Time Petri net with branching vertex

The reduced reachability graph will be obtained by repeatedly applying this procedure to all such paths.

The reduced reachability graph can also easily be constructed directly. An algorithm constructing the graph is given in Definition 3.39.

Definition 3.39 (reduced reachability graph) *Let* $\mathcal{Z} = (P, T, F, V, m_0, I)$ *with* $T = \{t_1, \cdots, t_n\}$ *be a finite Time Petri net. The reduced reachability graph of* $\mathcal{Z}$*, denoted by* $\mathcal{RG}_{\mathcal{Z}}^{redu}$*, is the directed labeled graph whose set of vertices* W*, set of edges* E *and edge labels from* $L \subseteq \mathbb{N} \times T$ *are defined by the following algorithm:*

begin $R := \{z_0\}$; $W := \emptyset$; $E := \emptyset$;

    while $R \neq \emptyset$ **do**

        Choose $z=(m,h)$ from R; $R := R - \{z\}$; $W := W \cup \{z\}$;

        if $\{t \in T \mid t^- \leq m\} \neq \emptyset$ **then**

            Let $\kappa := \min\{lft(t) - h(t) \mid t^- \leq m\}$;

            for $time = 0$ **to** κ **do**

                for i=1 **to** n **do**

                    if t_i ready to fire in $(m, h + time)$ **then**

                        Let z' be such that $z \xrightarrow{time} \xrightarrow{t} z'$;

                        $E := E \cup \{(z, [time, t], z')\}$;

                        if $z' \notin W$ **then** $R := R \cup \{z'\}$ **end**;

                end;

            end;

        end;

    end;

end.

In the following example we compute the reachability graph of the finite Time Petri net $\mathcal{Z}_5$ and afterwards reduce it as described in Definition 3.39. For the sake of simplicity we label edges with t instead of $0, t$.

Example 3.40

Let us consider the Time Petri net $\mathcal{Z}_5$, *given in Fig. 3.14. The reachability graph* $\mathcal{RG}_{\mathcal{Z}_5}$ *is computed according to Definition 3.38. The states* $z_0, \ldots, z_7$ *are the following:*

$$
\begin{array}{llll}
m_0 = (1,0,1,0) & h_1 = (\natural, 0, \natural) & h_5 = (1,1,\natural) & z_1 = (m_1, h_1) & z_5 = (m_0, h_5) \\
m_1 = (0,1,1,0) & h_2 = (\natural, 2, \natural) & h_6 = (\natural, 3, \natural) & z_2 = (m_1, h_2) & z_6 = (m_1, h_6) \\
m_2 = (0,1,0,1) & h_3 = (\natural, \natural, 0) & h_7 = (\natural, 1, \natural) & z_3 = (m_2, h_3) & z_7 = (m_1, h_7) \\
h_0 = (0,0,\natural) & h_4 = (\natural, \natural, 1) & z_0 = (m_0, h_0) & z_4 = (m_2, h_4) &
\end{array}
$$

$m'_n, h'_n)$ and $z^*_n = (m^*_n, h^*_n)$.

ved by induction on the length $\ell(\sigma)$ of σ:

s: $\ell(\sigma) = 0$

$\sigma = \varepsilon$ and $z'_n = z'_0 = z_0 = z^*_0 = z^*_n$. Thus the claim holds.

p: We assume the claim be true for any firing sequence $\sigma = t_1 \ldots t_n$ ngth n and will prove that it then also holds for the firing sequence $t_1 \ldots t_n t_{n+1}$.

let $\sigma(\tau)$ with $\tau = \tau_0 \ldots \tau_n$ be a run in $\mathcal{Z}$ feasible according to the inal rule. Let $w := t_1 \ldots t_n$ and $\kappa := \tau_0 \ldots \tau_{n-1}$. Then, σ and τ e the form $\sigma = w t_{n+1}$, $\tau = \kappa \tau_n$ and it holds that the following state nges in $\mathcal{Z}$ are feasible according to the original rule:

$$z_0 \xrightarrow{w(\kappa)} z'_n \xrightarrow{\tau_n} z''_n \xrightarrow{t_{n+1}} z'_{n+1}.$$

induction hypothesis, there also exists a run

$$z_0 \xrightarrow{w(\lambda)} z^*_n$$

ich is feasible in $\mathcal{Z}$ according to the modified rule and for which the im holds. We still have to prove that the run

$$z^*_n \xrightarrow{\theta_n} z^{**}_n \xrightarrow{t_{n+1}} z^*_{n+1} \qquad \text{with } \theta_n = \tau_n$$

feasible according to the modified rule and that the claim also holds r z'_{n+1} and z^*_{n+1}

et us consider the following two state changes in more detail:

$$z'_n \xrightarrow{\tau_n} z''_n \qquad \text{and} \tag{35}$$

$$z''_n \xrightarrow{t_{n+1}} z'_{n+1} \tag{36}$$

By the induction hypothesis it holds that $m'_n = m^*_n$. Therefore in he states z'_n and z^*_n the same transitions are enabled. Furthermore, or each transition t enabled in m'_n, (35) together with Definition 3.8 yields

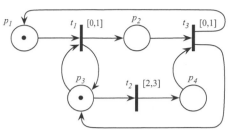

Figure 3.13: The Time Petri net $\mathcal{Z}_5$

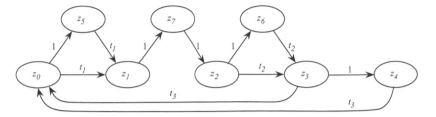

Figure 3.14: The reachability graph $\mathcal{RG}_{\mathcal{Z}_5}$

The reduced reachability graph $\mathcal{RG}^{redu}_{\mathcal{Z}_5}$ is as follows:

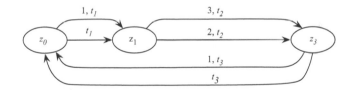

Figure 3.15: The reduced reachability graph $\mathcal{RG}^{redu}_{\mathcal{Z}_5}$

From now on we will not need to deal with the complete reachability graph of a finite Time Petri net $\mathcal{Z}$ and therefore we let $\mathcal{RG}_{\mathcal{Z}}$ denote the reduced reachability graph of $\mathcal{Z}$ for the rest of the book.

3.6 Reachability Graphs for Infinite Time Petri Nets

In this section we consistently extend the notion of reachability graphs to arbitrary Time Petri nets. The set of integer-states of an infinite Time Petri net can clearly be infinite even if the set of reachable p-markings of the net is finite. We will nevertheless show that for any Time Petri net there exists a finite set of reachable states, carrying all necessary information about the net properties (as for finite nets), if and only if the set of reachable p-markings in the net is finite. We will eventually see that if the net is finite this set of states is exactly the set of all reachable integer-states. This proves the consistency with the definition of integer-states for finite Time Petri nets.

We first modify Definition 3.8, which introduced the state change by elapsing time. For transitions with a finite lft nothing changes. For every transition t with $lft(t) = \infty$ we only consider the time until $eft(t)$. When t reaches $eft(t)$ we stop its clock $h(t)$ until t becomes disabled. We will prove that a p-marking in a Time Petri net is reachable using the modified definition if and only if it is reachable according to the original rule.

Definition 3.41 (modified state change) *Let* τ *be a non-negative real number and* $z = (m, h)$ *a state in the Time Petri net* $\mathcal{Z}$*. It is possible for time* τ *to elapse in the state* z *in* $\mathcal{Z}$ *if*

$$\forall t \, \big(\, t \in T \wedge h(t) \neq \sharp \longrightarrow h(t) + \tau \leq lft(t) \, \big).$$

The elapsing of time τ *will change* z *into the state* $z' = (m', h')$ *with*

1. $m' := m,$

2. $\forall t \, \Big(\, t \in T \longrightarrow h'(t) := \begin{cases} \sharp & \text{if } t^- \not\leq m' \\ eft(t) & \text{if } t^- \leq m' \, \wedge \\ & \qquad lft(t) = \infty \, \wedge \\ & \qquad eft(t) < h(t) + \tau \\ h(t) + \tau & \text{otherwise} \end{cases} \Big).$

Recall that $R_{\mathcal{Z}}$ denotes the set of all p-markings in a Time Petri net $\mathcal{Z}$ that are reachable according to Definitions 3.7 (firing) and 3.8 (elapsing of

time). We will write $R'_{\mathcal{Z}}$ for the set of a according to Definitions 3.7 and 3.41 (n

Theorem 3.42 *Let* $\mathcal{Z}$ *be a Time Pet marking* $\widetilde{m}$ *in* $\mathcal{Z}$ *that:*

$$\widetilde{m} \in R_{\mathcal{Z}} \text{ if and on}$$

Proof:

($\Longrightarrow$)

Let $\widetilde{m} \in R_{\mathcal{Z}}$. Then there exists a firing sec of σ which is feasible according to Definitio of state change) with $\tau = \tau_0 \tau_1 \ldots \tau_{n-1}$ and

$$z_0 \xrightarrow{\tau_0} z_0'' \xrightarrow{t_1} z_1' \xrightarrow{\tau_1} z_1'' \xrightarrow{t_2} z_2' \xrightarrow{\tau_2}$$

and $z_n' = (\widetilde{m}, h_n')$.

In order to prove the assertion it suffices t feasible according to Definitions 3.7 and 3.4 $\theta = \theta_0 \ldots \theta_{n-1}$ and

$$z_0 \xrightarrow{\theta_0} z_0^{**} \xrightarrow{t_1} z_1^* \xrightarrow{\theta_1} z_1^{**} \xrightarrow{t_2} z_2^* \xrightarrow{\theta_2} z_2^*$$

where it holds that $z_n^* = (\widetilde{m}, h_n^*)$. We wil stronger claim and thereby the necessity dired

Claim 3.43 *For each feasible run* $\sigma(\tau)$ *satisfy run* $\sigma(\theta)$ *satisfying (34) such that:*

(1) $\theta_{n-1} = \tau_{n-1},$

(2) $m_n^* = m_n',$

(3) *for all* $t \in T$ *it holds:* $h_n^*(t) = \begin{cases} eft(t) \\ \\ h_n'(t) & \end{cases}$

with $z_n' =$

This is pre

Bas

The

Ste
of l
$\sigma =$

So,
orig
hav
cha

By

w
c

is
f

L

$$h'_n(t) + \tau_n \leq lft(t).$$

Hence, starting in z_n^* time τ_n can also elapse according to the modified rule. We therefore set

$$\theta_n := \tau_n \tag{37}$$

and consider the state z_n^{**} with

$$z_n^* \xrightarrow{\theta_n} z_n^{**}.$$

It now holds that:

$$
\begin{aligned}
m''_n &= m'_n && \text{because of Def. 3.8}\\
&= m_n^* && \text{because of ind. hypothesis}\\
&= m_n^{**} && \text{because of Def. 3.41} \tag{38}
\end{aligned}
$$

In order to determine the t-marking h_{n+1}^* we first compute the preceding t-marking h_n^{**}. Let t be an arbitrary transition in T. We consider all possibilities for t:

Case 1: $t^- \not\leq m_n^{**}$.
Because of (38), this is true if and only if $t^- \not\leq m''_n$.
Hence, $h_n^{**}(t) = \sharp$ if and only if $h''_n(t) = \sharp$.

Case 2: $t^- \leq m_n^{**}$.

  Case 2.1: $lfl(t) = \infty$.

    Case 2.1.1: $h''_n(t) < eft(t)$.
    Then, because of

$$h''_n(t) = h'_n(t) + \tau_n,$$

it is also true that

$$h'_n(t) < eft(t)$$

which together with the induction hypothesis leads to

$$h_n^*(t) = h'_n(t). \tag{39}$$

But (39) and (37) then yield

$$\begin{aligned} h_n^*(t) + \theta_n &= h_n'(t) + \tau_n \\ &< \quad eft(t). \end{aligned} \qquad (40)$$

Thus, because of Definition 3.41 it holds that:

$$\begin{aligned} h_n^{**}(t) &= h_n^*(t) + \theta_n && \text{because of (40)} \\ &= h_n'(t) + \tau_n && \text{because of ind. hypo. and (37)} \\ &= h_n''(t). \end{aligned}$$

Case 2.1.2: $h_n''(t) \geq eft(t)$.

Case 2.1.2.1: $h_n'(t) < eft(t)$.
By induction hypothesis it follows that:

$$h_n^*(t) = h_n'(t).$$

Hence, it follows that

$$h_n^*(t) + \theta_n = h_n'(t) + \tau_n \geq eft(t).$$

Therefore, the modified rule in Definition 3.41 yields:

$$h_n^{**}(t) = eft(t).$$

Case 2.1.2.2: $h_n'(t) \geq eft(t)$.
Then, by induction hypothesis it holds that:

$$h_n^*(t) = eft(t)$$

and therefore the modified rule leads to

$$h_n^{**}(t) = eft(t).$$

Case 2.2: $lft(t) < \infty$.
Then, Definition 3.41 yields

$$\begin{aligned} h_n^{**}(t) &= h_n^*(t) + \theta_n && \text{because of ind. hypo. and (37)} \\ &= h_n'(t) + \tau_n \\ &= h_n''(t). \end{aligned}$$

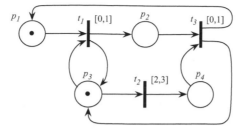

Figure 3.13: The Time Petri net $\mathcal{Z}_5$

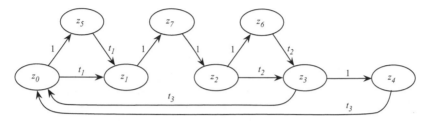

Figure 3.14: The reachability graph $\mathcal{RG}_{\mathcal{Z}_5}$

The reduced reachability graph $\mathcal{RG}^{redu}_{\mathcal{Z}_5}$ is as follows:

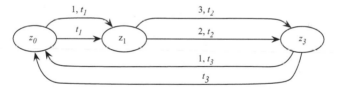

Figure 3.15: The reduced reachability graph $\mathcal{RG}^{redu}_{\mathcal{Z}_5}$

From now on we will not need to deal with the complete reachability graph of a finite Time Petri net $\mathcal{Z}$ and therefore we let $\mathcal{RG}_{\mathcal{Z}}$ denote the reduced reachability graph of $\mathcal{Z}$ for the rest of the book.

3.6 Reachability Graphs for Infinite Time Petri Nets

In this section we consistently extend the notion of reachability graphs to arbitrary Time Petri nets. The set of integer-states of an infinite Time Petri net can clearly be infinite even if the set of reachable p-markings of the net is finite. We will nevertheless show that for any Time Petri net there exists a finite set of reachable states, carrying all necessary information about the net properties (as for finite nets), if and only if the set of reachable p-markings in the net is finite. We will eventually see that if the net is finite this set of states is exactly the set of all reachable integer-states. This proves the consistency with the definition of integer-states for finite Time Petri nets.

We first modify Definition 3.8, which introduced the state change by elapsing time. For transitions with a finite lft nothing changes. For every transition t with $lft(t) = \infty$ we only consider the time until $eft(t)$. When t reaches $eft(t)$ we stop its clock $h(t)$ until t becomes disabled. We will prove that a p-marking in a Time Petri net is reachable using the modified definition if and only if it is reachable according to the original rule.

Definition 3.41 (modified state change) *Let* τ *be a non-negative real number and* $z = (m, h)$ *a state in the Time Petri net* $\mathcal{Z}$. *It is possible for time* τ *to elapse in the state* z *in* $\mathcal{Z}$ *if*

$$\forall t \left(t \in T \land h(t) \neq \sharp \longrightarrow h(t) + \tau \leq lft(t) \right).$$

The elapsing of time τ *will change* z *into the state* $z' = (m', h')$ *with*

1. $m' := m,$

2. $\forall t \left(t \in T \longrightarrow h'(t) := \begin{cases} \sharp & \text{if } t^- \not\leq m' \\ eft(t) & \text{if } t^- \leq m' \land \\ & \quad lft(t) = \infty \land \\ & \quad eft(t) < h(t) + \tau \\ h(t) + \tau & \text{otherwise} \end{cases} \right).$

Recall that $R_{\mathcal{Z}}$ denotes the set of all p-markings in a Time Petri net $\mathcal{Z}$ that are reachable according to Definitions 3.7 (firing) and 3.8 (elapsing of time). We will write $R'_{\mathcal{Z}}$ for the set of all p-markings in $\mathcal{Z}$ that are reachable according to Definitions 3.7 and 3.41 (modified state change).

Theorem 3.42 *Let $\mathcal{Z}$ be a Time Petri net. Then it holds for every p-marking $\tilde{m}$ in $\mathcal{Z}$ that:*

$$\tilde{m} \in R_{\mathcal{Z}} \text{ if and only if } \tilde{m} \in R'_{\mathcal{Z}}.$$

Proof:

($\Longrightarrow$)

Let $\tilde{m} \in R_{\mathcal{Z}}$. Then there exists a firing sequence $\sigma = t_1 \ldots t_n$ and a run $\sigma(\tau)$ of σ which is feasible according to Definitions 3.7 and 3.8 (original definitions of state change) with $\tau = \tau_0 \tau_1 \ldots \tau_{n-1}$ and

$$z_0 \xrightarrow{\tau_0} z_0'' \xrightarrow{t_1} z_1' \xrightarrow{\tau_1} z_1'' \xrightarrow{t_2} z_2' \xrightarrow{\tau_2} z_2'' \ldots \xrightarrow{\tau_{n-1}} z_{n-1}'' \xrightarrow{t_n} z_n' \qquad (33)$$

and $z_n' = (\tilde{m}, h_n')$.

In order to prove the assertion it suffices to show that there is a run $\sigma(\theta)$ feasible according to Definitions 3.7 and 3.41 (modified state change) with $\theta = \theta_0 \ldots \theta_{n-1}$ and

$$z_0 \xrightarrow{\theta_0} z_0^{**} \xrightarrow{t_1} z_1^* \xrightarrow{\theta_1} z_1^{**} \xrightarrow{t_2} z_2^* \xrightarrow{\theta_2} z_2^{**} \ldots \xrightarrow{\theta_{n-1}} z_{n-1}^{**} \xrightarrow{t_n} z_n^*, \qquad (34)$$

where it holds that $z_n^* = (\tilde{m}, h_n^*)$. We will actually prove the following stronger claim and thereby the necessity direction:

Claim 3.43 *For each feasible run $\sigma(\tau)$ satisfying (33) there exists a feasible run $\sigma(\theta)$ satisfying (34) such that:*

(1) $\theta_{n-1} = \tau_{n-1}$,

(2) $m_n^* = m_n'$,

(3) for all $t \in T$ it holds: $h_n^*(t) = \begin{cases} eft(t) & \text{if } h_n'(t) \neq \natural \wedge \\ & lft(t) = \infty \wedge \\ & eft(t) < h_n'(t) \\ h_n'(t) & \text{otherwise} \end{cases}$

with $z'_n = (m'_n, h'_n)$ and $z^*_n = (m^*_n, h^*_n)$.

This is proved by induction on the length $\ell(\sigma)$ of σ:

Basis: $\ell(\sigma) = 0$

Then $\sigma = \varepsilon$ and $z'_n = z'_0 = z_0 = z^*_0 = z^*_n$. Thus the claim holds.

Step: We assume the claim be true for any firing sequence $\sigma = t_1 \ldots t_n$ of length n and will prove that it then also holds for the firing sequence $\sigma = t_1 \ldots t_n t_{n+1}$.

So, let $\sigma(\tau)$ with $\tau = \tau_0 \ldots \tau_n$ be a run in $\mathcal{Z}$ feasible according to the original rule. Let $w := t_1 \ldots t_n$ and $\kappa := \tau_0 \ldots \tau_{n-1}$. Then, σ and τ have the form $\sigma = wt_{n+1}$, $\tau = \kappa\tau_n$ and it holds that the following state changes in $\mathcal{Z}$ are feasible according to the original rule:

$$z_0 \xrightarrow{w(\kappa)} z'_n \xrightarrow{\tau_n} z''_n \xrightarrow{t_{n+1}} z'_{n+1}.$$

By induction hypothesis, there also exists a run

$$z_0 \xrightarrow{w(\lambda)} z^*_n \qquad .$$

which is feasible in $\mathcal{Z}$ according to the modified rule and for which the claim holds. We still have to prove that the run

$$z^*_n \xrightarrow{\theta_n} z^{**}_n \xrightarrow{t_{n+1}} z^*_{n+1} \qquad \text{with } \theta_n = \tau_n$$

is feasible according to the modified rule and that the claim also holds for z'_{n+1} and z^*_{n+1}

Let us consider the following two state changes in more detail:

$$z'_n \xrightarrow{\tau_n} z''_n \qquad \text{and} \qquad (35)$$
$$z''_n \xrightarrow{t_{n+1}} z'_{n+1} \qquad\qquad\qquad (36)$$

By the induction hypothesis it holds that $m'_n = m^*_n$. Therefore in the states z'_n and z^*_n the same transitions are enabled. Furthermore, for each transition t enabled in m'_n, (35) together with Definition 3.8 yields

$$h'_n(t) + \tau_n \leq lft(t).$$

Hence, starting in z_n^* time τ_n can also elapse according to the modified rule. We therefore set

$$\theta_n := \tau_n \tag{37}$$

and consider the state z_n^{**} with

$$z_n^* \xrightarrow{\theta_n} z_n^{**}.$$

It now holds that:

$$
\begin{aligned}
m''_n &= m'_n \quad &\text{because of Def. 3.8} \\
&= m_n^* \quad &\text{because of ind. hypothesis} \\
&= m_n^{**} \quad &\text{because of Def. 3.41}
\end{aligned}
\tag{38}
$$

In order to determine the t-marking h_{n+1}^* we first compute the preceding t-marking h_n^{**}. Let t be an arbitrary transition in T. We consider all possibilities for t:

Case 1: $t^- \not\leq m_n^{**}$.

  Because of (38), this is true if and only if $t^- \not\leq m''_n$.
  Hence, $h_n^{**}(t) = \sharp$ if and only if $h''_n(t) = \sharp$.

Case 2: $t^- \leq m_n^{**}$.

  Case 2.1: $lft(t) = \infty$.

    Case 2.1.1: $h''_n(t) < eft(t)$.
      Then, because of

$$h''_n(t) = h'_n(t) + \tau_n,$$

it is also true that

$$h'_n(t) < eft(t)$$

which together with the induction hypothesis leads to

$$h_n^*(t) = h'_n(t). \tag{39}$$

But (39) and (37) then yield

$$h_n^*(t) + \theta_n \;=\; h_n'(t) + \tau_n$$
$$< \; eft(t). \tag{40}$$

Thus, because of Definition 3.41 it holds that:

$$h_n^{**}(t) \;=\; h_n^*(t) + \theta_n \qquad \text{because of (40)}$$
$$=\; h_n'(t) + \tau_n \qquad \text{because of ind. hypo. and (37)}$$
$$=\; h_n''(t).$$

Case 2.1.2: $h_n''(t) \geq eft(t)$.

Case 2.1.2.1: $h_n'(t) < eft(t)$.
By induction hypothesis it follows that:

$$h_n^*(t) = h_n'(t).$$

Hence, it follows that

$$h_n^*(t) + \theta_n = h_n'(t) + \tau_n \geq eft(t).$$

Therefore, the modified rule in Definition 3.41 yields:

$$h_n^{**}(t) = eft(t).$$

Case 2.1.2.2: $h_n'(t) \geq eft(t)$.
Then, by induction hypothesis it holds that:

$$h_n^*(t) = eft(t)$$

and therefore the modified rule leads to

$$h_n^{**}(t) = eft(t).$$

Case 2.2: $lft(t) < \infty$.
Then, Definition 3.41 yields

$$h_n^{**}(t) \;=\; h_n^*(t) + \theta_n \qquad \text{because of ind. hypo. and (37)}$$
$$=\; h_n'(t) + \tau_n$$
$$=\; h_n''(t).$$

Thus, the proof by cases for h_n^{**} is complete and the relationship between h_n^{**} and h_n'' is as follows:

$$h_n^{**}(t) = \begin{cases} eft(t) & \text{if } h_n''(t) \neq \sharp \wedge \\ & \qquad lft(t) = \infty \wedge \\ & \qquad eft(t) < h_n''(t) \\ h_n''(t) & \text{otherwise} \end{cases} \tag{41}$$

We now consider again the state change (36). It holds by Definition 3.7 that

$$t_{n+1}^- \leq m_n'' \qquad \text{and} \tag{42}$$
$$h_n''(t_{n+1}) \geq eft(t) \tag{43}$$

It follows from (42) together with (38) that

$$t_{n+1}^- \leq m_n^{**}$$

and from (43) and (41) that

$$eft(t_{n+1}) \leq h_n^{**}(t_{n+1}).$$

Thus, the transition t_{n+1} is ready to fire in the state z_n^{**}, i.e., there is a state $z_{n+1}^* = (m_{n+1}^*, h_{n+1}^*)$ in $\mathcal{Z}$ reachable with

$$z_n^{**} \xrightarrow{t_{n+1}} z_{n+1}^*.$$

It now holds for the p-markings m_{n+1}^* and m_{n+1}' that

$$\begin{aligned} m_{n+1}^* &= m_n^{**} + \Delta t_{n+1} & \text{by Definition 3.7} \\ &= m_n'' + \Delta t_{n+1} & \text{because of (36)} \\ &= m_{n+1}'. & \end{aligned} \tag{44}$$

We have thereby proved statement (2) of the claim.

For the t-marking h_{n+1}^* and an arbitrary transition $t \in T$ it holds that

Case 1: $t^- \not\leq m_{n+1}^*$.

Because of (44) it follows that

$$h_{n+1}^*(t) = \sharp - h_{n+1}'(t).$$

Case 2: $t^- \leq m_{n+1}^*$.

Case 2.1: $t^- \not\leq m_n^{**}$ or $\left(t^- \leq m_n^{**} \wedge {}^\bullet t \cap {}^\bullet t_{n+1} \neq \emptyset \right)$
In this case (38) leads to
$t^- \not\leq m_n''$ or $\left(t^- \leq m_n'' \wedge {}^\bullet t \cap {}^\bullet t_{n+1} \neq \emptyset \right)$,
and therefore it is true that
$h_{n+1}^*(t) = 0 = h_{n+1}'(t)$.

Case 2.2: $\left(t^- \leq m_n^{**} \wedge {}^\bullet t \cap {}^\bullet t_{n+1} = \emptyset \right)$
Then, because of Definition 3.7 it holds that:

$$
\begin{aligned}
h_{n+1}^*(t) &= h_n^{**}(t) \qquad \text{and} \\
h_{n+1}'(t) &= h_n''(t).
\end{aligned}
$$

Hence, because of (41) statement (3) of the claim follows and the proof of the claim is compete.

$(\Longleftarrow)$

Let $\tilde{m} \in R_z'$. Then there is a transition sequence σ and a run $\sigma(\theta)$ of σ feasible according to the modified rule, such as in (34). It now suffices to show that there is also a run $\sigma(\tau)$ of σ feasible according to the original rule, such as in (33). We now state the claim which is "dual" to Claim 3.43 and can be proved similarly exchanging *-notations for '-notations and '-notations for *-notations.

Claim 3.44 *For each feasible run $\sigma(\theta)$ satisfying (34) there exists a feasible run $\sigma(\tau)$ satisfying (33) such that*

(1) $\tau_{n-1} = \theta_{n-1}$,

(2) $m_n' = m_n^*$,

(3) *for all $t \in T$ it holds:*
$$
\begin{cases}
h_n'(t) \geq eft(t) & \text{if } h_n^*(t) \neq \sharp \wedge \\
& \quad lft(t) = \infty \wedge \\
& \quad eft(t) = h_n^*(t) \\
h_n'(t) = h_n^*(t) & \text{otherwise}
\end{cases}
$$

with $z_n' = (m_n', h_n')$ and $z_n^* = (m_n^*, h_n^*)$.

Claim 3.44 immediately yields the sufficiency direction for Theorem 3.42. □

The following corollary is a direct consequence of theorem 3.42.

Corollary 3.45 *For any Time Petri net $\mathcal{Z}$ it holds that $R_{\mathcal{Z}} = R'_{\mathcal{Z}}$.*

Definition 3.46 (essential-state) *An integer-state $z = (m, h)$ in a Time Petri net $\mathcal{Z}$ is called essential-state when $\mathcal{Z}$ is defined with the modified firing rule.*

It is obvious that for an arbitrary essential-state $z = (m, h)$ and for each transition t with $lft(t) = \infty$ in a Time Petri net $\mathcal{Z}$ the clock $h(t)$ has at most the value $eft(t)$.

Similar to the set $RIS_{\mathcal{Z}}$ of all integer-states reachable in a Time Petri net $\mathcal{Z}$ according to the original rule, we consider the set $REIS_{\mathcal{Z}}$ of all integer-states reachable according to the modified rule, and call these states *reachable essential-states*.

Definition 3.47 (reachable essential-states) *Let $\mathcal{Z}$ be an arbitrary Time Petri net. The set $REIS_{\mathcal{Z}}$ of all reachable essential-states in $\mathcal{Z}$ is defined as follows:*

$$REIS_{\mathcal{Z}} := \{ z \mid z_0 \xrightarrow{\sigma(\tau)} z, \ z \text{ is an essential-state and}$$

$$\sigma(\tau) \text{ is a run feasible in } Z, \text{ according to}$$

$$\text{Definitions 3.7 and 3.41} \}.$$

We call the states in $REIS_{\mathcal{Z}}$ *essential* because they carry enough information to decide almost all net properties of interest, such as reachability of p-markings, liveness, reversibility etc.

It is obvious that for every finite Time Petri net $\mathcal{Z}$ it holds that $REIS_{\mathcal{Z}} = RIS_{\mathcal{Z}}$. We will see in the next example that in an infinite Time Petri net neither of these sets is necessarily contained in the other.

Example 3.48

Let us consider the Time Petri net $\mathcal{Z}$:

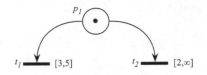

Figure 3.16:   The infinite Time Petri net $\mathcal{Z}$

It holds that:
$$RIS_{\mathcal{Z}} = \{(1, \begin{pmatrix} 0 \\ 0 \end{pmatrix}), (1, \begin{pmatrix} 1 \\ 1 \end{pmatrix}), (1, \begin{pmatrix} 2 \\ 2 \end{pmatrix}), (1, \begin{pmatrix} 3 \\ 3 \end{pmatrix}), (1, \begin{pmatrix} 4 \\ 4 \end{pmatrix}), (1, \begin{pmatrix} 5 \\ 5 \end{pmatrix}),$$
$$(0, \begin{pmatrix} \sharp \\ \sharp \end{pmatrix})\} \ and$$
$$REIS_{\mathcal{Z}} = \{(1, \begin{pmatrix} 0 \\ 0 \end{pmatrix}), (1, \begin{pmatrix} 1 \\ 1 \end{pmatrix}), (1, \begin{pmatrix} 2 \\ 2 \end{pmatrix}), (1, \begin{pmatrix} 3 \\ 2 \end{pmatrix}), (1, \begin{pmatrix} 4 \\ 2 \end{pmatrix}), (1, \begin{pmatrix} 5 \\ 2 \end{pmatrix}),$$
$$(0, \begin{pmatrix} \sharp \\ \sharp \end{pmatrix})\}.$$

Clearly, neither $REIS_{\mathcal{Z}} \subseteq RIS_{\mathcal{Z}}$ *nor* $REIS_{\mathcal{Z}} \supseteq RIS_{\mathcal{Z}}$.

Furthermore, Theorems 3.26 and 3.32 can also be proved using the modified rule (Definition 3.41 instead of Definition 3.8) for elapsing time. This means that every reachable p-marking in $\mathcal{Z}$ can be reached with a feasible run where all elapsing times are non-negative integers and all states reached during the run are essential-states.

Moreover, for the set $REIS_{\mathcal{Z}}$ the following generalization of Theorem 3.37 holds:

Theorem 3.49 *Let* $\mathcal{Z}$ *be an arbitrary Time Petri net.   Then,* $REIS_{\mathcal{Z}}$ *is finite if and only if the set* $R_{\mathcal{Z}}$ *is finite.*

Proof:

$(\Longrightarrow)$

Let $REIS_{\mathcal{Z}}$ be finite. Hence, the set $R'_{\mathcal{Z}}$ of all p-markings reachable in $\mathcal{Z}$ according to the modified rule is also finite. This together with Corollary 3.45 shows that $R_{\mathcal{Z}}$ is finite.

$(\Longleftarrow)$

Let $R_{\mathcal{Z}}$ be finite. Because of Corollary 3.45 the set $R'_{\mathcal{Z}}$ is also finite.

Let $\widetilde{m} \in R'_{\mathcal{Z}}$. The set of all possible integer-states suitable for $\widetilde{m}$ (reachable as well as not reachable according to the modified rule) is obviously finite.

$$
\begin{aligned}
A_{\widetilde{m}} := \{ z \mid z = (\widetilde{m}, h) \wedge & \\
\forall t \big((t \in T \wedge t^- \not\leq \widetilde{m}) &\longrightarrow h(t) = \natural \big) \wedge \\
\forall t \big((t \in T \wedge t^- \leq \widetilde{m}) &\longrightarrow \big(h(t) \in \mathbb{N} \wedge \\
\big((lft(t) < \infty &\longrightarrow 0 \leq h(t) \leq lft(t)) \vee \\
(lft(t) = \infty &\longrightarrow 0 \leq h(t) \leq eft(t)) \big) \big) \big) \}
\end{aligned}
$$

But since

$$
REIS_{\mathcal{Z}} \subseteq \bigcup_{\widetilde{m} \in R'_{\mathcal{Z}}} A_{\widetilde{m}}
$$

it follows that $REIS_{\mathcal{Z}}$ is also finite. $\qquad\square$

Based on the modified rule for state change from Definition 3.41 we can now consistently extend the definition of a reachability graph to arbitrary Time Petri nets. To reduce the resulting reachability graph we adapt the algorithm from Definition 3.39.

Definition 3.50 (reachability graph for arbitrary Time Petri nets)
Let $\mathcal{Z} = (P, T, F, V, m_0, I)$ be an arbitrary (finite or infinite) Time Pctri net with $T = \{t_1, \cdots, t_n\}$. The (reduced) reachability graph $\mathcal{RG}^{redu}_{\mathcal{Z}} := (W, E, L)$ of $\mathcal{Z}$ is the directed graph with edge labels whose set of vertices W, set of edges E and edge labels from $L \subseteq \mathbb{N} \times T$ are defined by the following algorithm:

begin $R := \{z_0\}; \quad W := \emptyset; \quad E := \emptyset;$

    while $R \neq \emptyset$ **do**

        Choose $z=(m,h)$ from $R; \quad R := R - \{z\}; \quad W := W \cup \{z\};$

        if $\{t \in T \mid t^- \leq m\} \neq \emptyset$ **then**

            if $\{t \in T \mid t^- < m \wedge lft(t) \neq \infty\} \neq \emptyset$

then Let $\kappa := \min\{lft(t) - h(t) \mid t^- \leq m\}$

else Let $\kappa := \max\{eft(t) - h(t) \mid t^- \leq m\}$

end;

for $time = 0$ to κ do

    for i=1 to n do

        if t_i ready to fire in $(m, h + time)$ then

            Let z' be such that $z \xrightarrow{time} \xrightarrow{t_i} z'$;

            $E := E \cup \{(z, [time, t_i], z')\}$;

            if $z' \notin W$ then $R := R \cup \{z'\}$ end;

        end;

    end;

end;

end;

end.

Example 3.51 *We consider the infinite Time Petri net $\mathcal{Z}_6$, obtained from the finite Time Petri net $\mathcal{Z}_5$ from Example 3.40 by changing the interval of transition t_2 to $[2, \infty]$:*

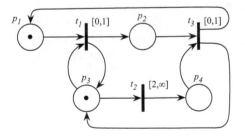

Figure 3.17:   The Time Petri net $\mathcal{Z}_6$

The reachability graph $\mathcal{RG}_{\mathcal{Z}_6}$ computed using Definition 3.38, Definition 3.41 and Definition 3.7 is represented in Fig. 3.18. The states z_0, z_1 and z_3 are

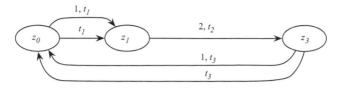

Figure 3.18: The reachability graph $\mathcal{RG}_{\mathcal{Z}_6}$

the same as in example 3.40.

$z_0 = ((1,0,1,0),((0,0,\sharp)), z_1 = ((0,1,1,0),(\sharp,0,\sharp)), z_3 = ((0,1,0,1),(\sharp,\sharp,0)).$

3.7 Qualitative Properties

In this section we introduce reachability and liveness in Time Petri nets and study these basic properties in detail. As already mentioned, they are essential for a comprehensive analysis of a net. The definitions of the properties should of course be consistent with those for classic Petri nets, meaning that for a Time Petri net where every transition is assigned the time interval $[0, \infty]$ the respective notions should agree.

3.7.1 Reachability

In Section 3.2 we introduced *reachability of states* and *p-markings*. In Section 3.3 we showed that Time Petri nets and counter machines and thus, Time Petri nets and Turing machines have the same computational power. To prove this we modeled natural numbers with *p*-markings and saw that deciding reachability of *p*-markings in Time Petri nets is as hard as the halting problem for Turing machines and therefore not decidable.

Questions about the reachability of a *p*-marking are actually of great practical relevance because the *p*-markings model possible situations in a system.

Another question that can be important for a time-dependent system is under which circumstances of time a situation can occur. We will study such quantitative questions in the next section.

It is generally not decidable for a p-marking in an arbitrary Time Petri net whether it is reachable. Requiring additional conditions of the considered nets can however make answering the question possible.

In the following we systematize results about the reachability of p-markings and states in Time Petri nets, considering different kinds of states and different classes of nets. Let $\mathcal{Z}$ be an arbitrary Time Petri net and $S(\mathcal{Z})$ its skeleton. We will call a state $z = (m, h)$ a *rational-state* if for every transition t in $\mathcal{Z}$ it holds that if $h(t) \neq \natural$ then $h(t) \in \mathbb{Q}_0^+$. As already mentioned, the set of all reachable p-markings in $\mathcal{Z}$ is a subset of the reachable markings in its skeleton $S(\mathcal{Z})$.

Proposition 3.52 *Let the skeleton $S(\mathcal{Z})$ of the Time Petri net $\mathcal{Z}$ be bounded. Then it holds that:*

1. *The reachability of any p-marking m^* in $\mathcal{Z}$ is decidable.*

2. *The reachability of any rational-state z^* in $\mathcal{Z}$ is decidable.*

Proof:

For 1.

Let $S(\mathcal{Z})$ be bounded and let m^* be an arbitrary p-marking in $\mathcal{Z}$. Then, $R_{S(\mathcal{Z})}$ is finite. Hence, by Proposition 3.37 it follows that the set of all reachable essential-states $REIS_{\mathcal{Z}}$ is also finite. But because of Remark 3.30[7] there exists for each reachable p-marking m in $\mathcal{Z}$ a reachable essential-state whose p-marking is m. Thus, in order to decide whether m^* is reachable in $\mathcal{Z}$ we can simply check whether the finite set $REIS_{\mathcal{Z}}$ contains a state whose p-marking is m^*.

For 2.

Let $S(\mathcal{Z})$ be bounded and let $z^* = (m^*, h^*)$ be an arbitrary rational-state. Let $h^*(t) = \frac{r_t}{q_t}$ with $r_t, q_t \in \mathbb{N}$ for all $t \in T$ with $h^*(t) \neq \natural$. We define λ as

---

[7]Remark 3.30 was proved using the original rule for state change. It is easy to see that the remark also holds using the modified rule for state change.

follows:

$$\lambda := LCM\{q_t \mid t \in T \wedge h^*(t) \neq \natural\}.$$

We now scale down the measuring unit for time in $\mathcal{Z}$ by the factor λ, i.e., we multiply all interval bounds in the net by λ. Hence, $h^*(t)$ in the new measuring unit is λ times larger than in the original measuring unit and therefore after the transformation $h^*(t) \in \mathbb{N}$ for all $t \in T$ with $h^*(t) \neq \natural$. Thus, the state z^* is an integer-state (essential) with respect to the new measuring unit. As $R_{S(\mathcal{Z})}$ is finite and z^* is an integer-state the decidability of z^* follows by Proposition 3.37. $\square$

We will call a state $z = (m, h)$ in $\mathcal{Z}$ a *proper real-state* if there is at least one transition t in $\mathcal{Z}$ such that $h(t) \in \mathbb{R}_0^+ \setminus \mathbb{Q}_0^+$. The reachability of such states is in general not decidable without additional information. Remark 3.35 and Corollary 3.34 give us the following sufficient condition for non-reachability of an arbitrary state:

Proposition 3.53 *For any state z in the Time Petri net $\mathcal{Z}$ it holds that if $\underline{z}$ or $\overline{z}$ is not reachable in $\mathcal{Z}$ then z is not reachable in $\mathcal{Z}$ either.*

Under additional restrictions for the Time Petri net, the reachability of proper real-states can be decidable. In order to discuss some such classes of nets we first prove the following property for every Time Petri net.

Proposition 3.54 *Let $\mathcal{Z}$ be a Time Petri net and m^* a p-marking reachable in $S(\mathcal{Z})$ by firing the sequence σ. Furthermore let z^* be a state with p-marking m^*. Then it is decidable whether there is a feasible run $\sigma(\tau)$ in $\mathcal{Z}$ such that:*
$$z_0 \xrightarrow{\sigma(\tau)} z^*.$$

Proof:

Let us consider the parametric state $\left(z_\sigma = (m_\sigma, h_\sigma), B_\sigma\right)$ in $\mathcal{Z}$. Clearly, if B_σ has at least one solution it holds that $m_\sigma = m^*$. We consider the system of linear inequalities

$$\begin{cases} B_\sigma \\ h_\sigma(t) = h^*(t) \text{ for every } t \in T. \end{cases}$$

If the system is solvable (in the field of non-negative real numbers), then each solution gives us a feasible run of σ such that after firing this run the state z^* is reached. If the system is not solvable, then there is no feasible run of σ with the desired property. The solvability of a system of linear inequalities is, of course, decidable in polynomial time (c.f. [PS98], [GLS93]), and therefore the reachability of a state is decidable if its p-marking is reachable.   □

Let us again consider an arbitrary bounded Time Petri net. The reachability graph of such a net is, according to Theorem 3.37, a finite, directed, weighted graph. Let $z^* = (m^*, h^*)$ be a proper real-state. If there is no reachable integer-state with the p-marking m^*, then z^* is not reachable. But if there is such a reachable integer-state z_{m^*} then there also is a path from z_0 to z_{m^*} in the reachability graph. Each path in the reachability graph corresponds to a feasible run of some firing sequence σ. According to Proposition 3.54 we can decide for any given finite firing sequence whether there is a feasible run of σ by which the state z^* is reached. Note that it is still undecidable for an arbitrary state in a bounded Time Petri net whether there exists a run by which this state is reachable.

Let us now consider unbounded Time Petri nets. The skeletons of such nets are clearly also unbounded and the inclusion $R_Z \subseteq R_{S(Z)}$ holds. Thus, when a p-marking is not reachable in the skeleton (and this is decidable) this p-marking can not be reached in Z either. As shown in Section 3.3, it is in general not decidable whether a p-marking is reachable in an arbitrary Time Petri net, but we have also mentioned that with additional restrictions on the net this question can be answered. We will consider two restricted classes of Time Petri nets for which time does not affect the reachability of p-markings. Since the reachability of markings in a timeless net is decidable, c.f. [PW03], the reachability of p-markings, as well, is decidable for these classes.

We will now consider first "speeded" and then "lazy" Time Petri nets. A *speeded net* is a Time Petri net such that $eft(t) = 0$ for each transition t in the net and a *lazy net* is a Time Petri net with $lft(t) = \infty$ for each transition t in the net.

Proposition 3.55 (speeded nets and reachability) *Let $Z = (P, T, F, V, m_0, I)$ be a Time Petri net. If every transition in T is ready to fire immediately after being enabled then the set of all reachable p-markings in Z is the same as the set of all reachable markings in its skeleton. Formally:*

$$\bigl(\forall t\,(\,t \in T \longrightarrow eft(t) = 0\,)\ \longrightarrow\ R_{\mathcal{Z}} = R_{S(\mathcal{Z})}\bigr). \tag{45}$$

Proof: In order to show the equality between the sets $R_{\mathcal{Z}}$ and $R_{S(\mathcal{Z})}$ we prove the inclusions $R_{\mathcal{Z}} \subseteq R_{S(\mathcal{Z})}$ and $R_{S(\mathcal{Z})} \subseteq R_{\mathcal{Z}}$.

$(R_{\mathcal{Z}} \subseteq R_{S(\mathcal{Z})})$:

This inclusion holds for all Time Petri nets.

$(R_{S(\mathcal{Z})} \subseteq R_{\mathcal{Z}})$:

Let $m \in R_{S(\mathcal{Z})}$. Then there is a firing sequence σ in the skeleton $S(\mathcal{Z})$ of $\mathcal{Z}$ with

$$m_0 \xrightarrow{\ \sigma\ } m.$$

Now we consider the run $\sigma(\tau)$ of σ in the net $\mathcal{Z}$ with $\tau = \tau_0\tau_1 \dots \tau_{\ell(\sigma)}$ and $\tau_i = 0$ for all $i, 0 \le i \le \ell(\sigma)$. Because of the premise in (45), this run is feasible in $\mathcal{Z}$. The full proof can be done by induction on $\ell(\sigma)$. $\qquad\square$

Proposition 3.56 (lazy nets and reachability) *Let $\mathcal{Z} = (P, T, F, V, m_0, I)$ be a Petri net. If no transition in T is ever forced to fire then the set of p-markings reachable in $\mathcal{Z}$ is the same as the set of markings reachable in its skeleton. Formally:*

$$\bigl(\forall t\,(\,t \in T \longrightarrow lft(t) = \infty\,)\ \longrightarrow\ R_{\mathcal{Z}} = R_{S(\mathcal{Z})}\bigr). \tag{46}$$

Proof: In order to show the equality between $R_{\mathcal{Z}}$ and $R_{S(\mathcal{Z})}$ we again prove the inclusion $R_{S(\mathcal{Z})} \subseteq R_{\mathcal{Z}}$, as $R_{\mathcal{Z}} \subseteq R_{S(\mathcal{Z})}$ holds for all Time Petri nets.

$(R_{S(\mathcal{Z})} \subseteq R_{\mathcal{Z}})$:

Let $m \in R_{S(\mathcal{Z})}$. Then there is a firing sequence σ in the skeleton $S(\mathcal{Z})$ of $\mathcal{Z}$ with

$$m_0 \dashrightarrow^{\sigma} m.$$

It is easy to see that, because of the premise of (46), the run

$$z_0 \xrightarrow{\ \tau_0\ } z_0' \xrightarrow{\ t_1\ } z_1 \dots \xrightarrow{\ \tau_{n-1}\ } z_{n-1}' \xrightarrow{\ t_n\ } z_n$$

with $z_i = (m_i, h_i)$ and $z_i' = (m_i', h_i')$, where $m_n = m$ and $\tau_i := \max\{eft(t) \mid t \in T\}$ for all $i = 0, \dots, n-1$ is feasible in $\mathcal{Z}$. The full proof can be done by induction on $\ell(\sigma)$. $\square$

In unbounded Time Petri nets the reachability of p-markings and of rational-states is semi-decidable. Propositions 3.53 and 3.54 hold for all Time Petri nets and give sufficient conditions for the non-reachability of arbitrary states.

3.7.2   Liveness

In this section we will consistently extend the notion of *liveness*, which we have defined for timeless Petri nets, to Time Petri nets. There is not as direct a relationship between the liveness of a Time Petri net and the liveness of its skeleton as for reachability. For example it is possible that a transition which is enabled in the skeleton can never become enabled in the Time Petri net, or if enabled in the time-dependent net the transition still might never become ready to fire. In some special cases there is however a close relationship between the liveness of the two nets.

With the next definition we introduce the property of *liveness* for arbitrary Time Petri nets.

Definition 3.57 (liveness) *Let* $\mathcal{Z} = (P, T, F, V, m_0, I)$ *be a Time Petri net, z a reachable state in $\mathcal{Z}$ and $t \in T$. Then*

1. *t is live in the state z in $\mathcal{Z}$ if for every state $z' \in RS_{\mathcal{Z}}(z)$ there exists a state $z'' \in RS_{\mathcal{Z}}(z')$ such that: $z'' \xrightarrow{t}$.*

2. *t is dead in z in $\mathcal{Z}$ if for all states $z' \in RS_{\mathcal{Z}}(z)$ transition t is not ready to fire in z'.*

3. *z is live in $\mathcal{Z}$ if all transitions $t \in T$ are live in z.*

4. *z is dead in $\mathcal{Z}$ if all transitions $t \in T$ are dead in z.*

5. *t is live/dead in $\mathcal{Z}$ if t is live/dead in z_0.*

6. *$\mathcal{Z}$ is live/dead if z_0 is live/dead in $\mathcal{Z}$.*

7. *$\mathcal{Z}$ is blocking-free if there is no reachable state in $\mathcal{Z}$ such that all transitions are dead in this state.*

The following example illustrates that information about liveness in the skeleton of a Time Petri net in general does not tell us anything about liveness in the time-dependent net, nor the other way around.

Example 3.58 *Let us consider the Time Petri nets $\mathcal{Z}_5$ and $\mathcal{Z}_6$ shown in Fig. 3.19 and Fig. 3.20:*

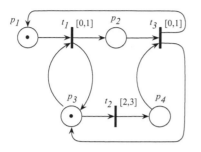
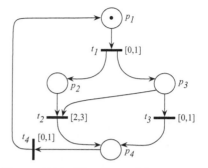

Figure 3.19: The Time Petri net $\mathcal{Z}_5$ Figure 3.20: The Time Petri net $\mathcal{Z}_6$

- *The transition t_2 is live in z_0 in $\mathcal{Z}_5$ but t_2 is not live in m_0 in $S(\mathcal{Z}_5)$.*

- *t_2 is not live in z_0 in $\mathcal{Z}_6$ (it is even dead) but t_2 is live in m_0 in $S(\mathcal{Z}_6)$.*

- *t_2 is live in $\mathcal{Z}_5$ but t_2 is not live in $S(\mathcal{Z}_5)$.*

- *t_2 is not live in $\mathcal{Z}_6$ (it is even dead), but t_2 is live in $S(\mathcal{Z}_6)$.*

- *z_o is live in $\mathcal{Z}_5$ but m_o is not live in $S(\mathcal{Z}_5)$.*

- *z_o is not live in $\mathcal{Z}_6$ but m_o is live in $S(\mathcal{Z}_6)$.*

- *$\mathcal{Z}_5$ is live but $S(\mathcal{Z}_5)$ is not live.*

- *$\mathcal{Z}_6$ is not live but $S(\mathcal{Z}_6)$ is live.*

For some restricted classes of Time Petri nets the time has no effect on liveness at all. For such classes the time-dependent net is live if its skeleton is live. For speeded and lazy nets the time-dependent net, moreover, is live if and only if its skeleton is live. With Propositions 3.55 and 3.56 we have shown that the set of reachable p-markings of a speeded or lazy Time Petri net is the same as the set of reachable markings of its skeleton. Hence, in such a net a transition is enabled if and only if it is enabled in the skeleton. However this does not mean that all enabled transitions can become ready to fire in the time-dependent net.

Proposition 3.59 (speeded nets and liveness) *Let $\mathcal{Z} = (P, T, F, V, m_0, I)$ be a Time Petri net. If every transition in T is ready to fire immediately after its enabling then $\mathcal{Z}$ is live if and only if its skeleton $S(\mathcal{Z})$ is live. Formally:*

$$\big(\forall t \, (t \in T \longrightarrow eft(t) = 0 \,) \ \longrightarrow \ (\mathcal{Z} \text{ is live} \longleftrightarrow S(\mathcal{Z}) \text{ is live}) \big).$$

Proof:

Let $\mathcal{Z}$ be a speeded Time Petri net, i.e., for each transition $t \in T$ it holds that $eft(t) = 0$. We will prove the two directions of the proposition separately:

($\Longrightarrow$)

We will first show that if $\mathcal{Z}$ is live then $S(\mathcal{Z})$ is live, too.

Let $\mathcal{Z}$ be live. Let $t \in T$ and $m' \in R_{S(\mathcal{Z})}$. By Proposition 3.55 it holds that: $R_{S(\mathcal{Z})} = R_{\mathcal{Z}}$. Hence, m' is a reachable p-marking in $\mathcal{Z}$, and therefore there exits a state $z' = (m', h') \in RS_{\mathcal{Z}}$. But, because $\mathcal{Z}$ is live there exists a state $z'' = (m'', h'') \in RS_{\mathcal{Z}}(z')$ with $z'' \xrightarrow{t}$, i.e., t is enabled in m''.

Let us examine the p-marking m'' more closely: Because z'' is reachable in $\mathcal{Z}$ from z' it holds by definition that m'' is reachable in $\mathcal{Z}$ from m', i.e., $m'' \in R_{\mathcal{Z}}(m')$. Furthermore, according to Proposition 3.55 it holds that $R_{S(\mathcal{Z})} = R_{\mathcal{Z}}$ and hence, it immediately follows that $m'' \in R_{S(\mathcal{Z})}(m')$. We know that t is enabled in m'', i.e., $m'' \xrightarrow{t}$.

We have shown that there is for each transition t and each reachable marking m' in $S(\mathcal{Z})$ a further marking m'' reachable from m' in which t is enabled. Therefore, by Definition 2.21 the classic Petri net $S(\mathcal{Z})$ is live, too.

($\Longleftarrow$)

We now prove that if $S(\mathcal{Z})$ is live then $\mathcal{Z}$ is live, too.

Let $S(\mathcal{Z})$ be live. Let $t \in T$ and $z' = (m', h') \in RS_{\mathcal{Z}}$. Because z' is reachable in $\mathcal{Z}$ the p-marking m' is by Definition 3.13 reachable in $\mathcal{Z}$, i.e., $m' \in R_{\mathcal{Z}}$. Because of Proposition 3.55 it then holds that the p-marking m' is also reachable in the skeleton of $\mathcal{Z}$, i.e., $m' \in R_{S(\mathcal{Z})}$. Now, because of the liveness of the skeleton and by Definition 2.21 (liveness in timeless nets) it is always possible to reach a further marking m'' from the marking m' such that t is enabled in m'', i.e.,

$$\exists m'' \left(m'' \in R_{S(\mathcal{Z})}(m') \wedge m'' \xrightarrow{t} \right)$$

or in other words, there is a firing sequence σ with

$$m' \xrightarrow{\sigma} m'' \xrightarrow{t} . \tag{47}$$

Let us consider the run $\sigma(\tau)$ in $\mathcal{Z}$ with $\tau := \tau_0 \tau_1 \ldots \tau_{\ell(\sigma)-1}$ and $\tau_i := 0$ for $i = 0, \ldots, \ell(\sigma) - 1$. By assumption the net $\mathcal{Z}$ is a speeded one, i.e., all transitions in $\mathcal{Z}$ can fire immediately after their enabling. Hence, the run $\sigma(\tau)$ is a feasible one, i.e.,

$$z' \xrightarrow{\sigma(\tau)} \text{ is feasible in } \mathcal{Z}.$$

Thus, from z' a state $z'' = (m'', h'')$ will be reached in $\mathcal{Z}$, i.e., $z' \xrightarrow{\sigma(\tau)} z''$ and therefore $z'' \in RS_{\mathcal{Z}}(z')$.

Because t is enabled in m'' and $\mathcal{Z}$ is a speeded net t is ready to fire in z''. Consequently and by Definition 3.57 the Time Petri net $\mathcal{Z}$ is live, too. $\square$

Proposition 3.60 (lazy nets and liveness) *Let $\mathcal{Z} = (P, T, F, V, m_0, I)$ be a Time Petri net. If no transition in T is ever forced to fire then $\mathcal{Z}$ is live if and only if its skeleton $S(\mathcal{Z})$ is live.*
Formally:

$$\left(\forall t \, (t \in T \longrightarrow lft(t) = \infty) \longrightarrow (\mathcal{Z} \text{ is live} \longleftrightarrow S(\mathcal{Z}) \text{ is live}) \right).$$

Idea of the proof:

Let $\mathcal{Z}$ be a lazy net. Again and similar to proposition 3.59, we will prove the two directions of the proposition separately

$(\Longrightarrow)$

Let us assume that $\mathcal{Z}$ is live and show that $S(\mathcal{Z})$ is live, too.

By Proposition 3.56 it holds for every lazy Time Petri net that the set of all its reachable p-markings is the same as the set of all markings reachable in its skeleton. Similarly as in the proof of Proposition 3.59 we conclude that each firing sequence in the lazy Time Petri net. is also a firing sequence in the skeleton. Therefore $S(\mathcal{Z})$ is live.

$(\Longleftarrow)$

Assuming that $S(\mathcal{Z})$ is live we will show that $\mathcal{Z}$ is also live.

This direction of the proof is also similar to the respective direction of the proof of Proposition 3.59. For every firing sequence σ in the skeleton the run $\sigma(\tau)$ with $\tau_i := \max\{eft(t) \mid t \in T\}$ for all $i = 1, \ldots, \ell(\sigma)$ is a feasible one. The time elapsing between two firings ensures that in every reachable p-marking in $\mathcal{Z}$ all enabled transitions become ready to fire before one of them fires. $\square$

With the next theorem we will prove for a further class of Time Petri nets that they are live if their skeleton is live. Thus, we present a third subset of Time Petri nets which is time-invariant with respect to liveness. Please note that all additional constraints for this subset are statical, and therefore decidable. It should also be noted that unlike for speeded and lazy nets $R_{\mathcal{Z}} = R_{S(\mathcal{Z})}$ does not necessarily hold for every net $\mathcal{Z}$ in the class of Time Petri nets introduced in the next theorem. We will later give an example of a net Z from this class for which $R_{\mathcal{Z}} \neq R_{S(\mathcal{Z})}$ (cf. Remark 3.62).

The following two notations will be used in the theorem. Let $\mathcal{Z} = (P, T, F, V, m_0, I)$ be a Time Petri net and p a place in $\mathcal{Z}$. Then

$$\mathcal{M}in(p) := \max\{\, eft(t) \mid t \in T \,\wedge\, t \in p^{\bullet} \,\}$$

and

$$\mathcal{M}ax(p) := \min\{\, lft(t) \mid t \in T \,\wedge\, t \in p^{\bullet} \,\}.$$

Theorem 3.61 *Let* $\mathcal{Z} = (P, T, F, V, m_0, I)$ *be a Time Petri net such that* *(a)* $S(\mathcal{Z})$ *is an EFC net*,[8]

(b) $S(\mathcal{Z})$ *is homogeneous,*

(c) for each place $p \in P$ *it holds that* $\mathcal{M}in(p) \leq \mathcal{M}ax(p)$ *and*

(d) for each transition $t \in T$ *it holds that* $lft(t) > 0$.

Then if $S(\mathcal{Z})$ *is live,* $\mathcal{Z}$ *is live, too.*

Proof:

Let the skeleton $S(\mathcal{Z})$ of the Time Petri net $\mathcal{Z}$ be live. Furthermore let $z' = (m', h') \in RS_{\mathcal{Z}}$ and $t^* \in T$.

In order to prove that $\mathcal{Z}$ is also live it is sufficient to show that there is a state $z'' \in R_{\mathcal{Z}}(z')$ such that firing of $z'' \xrightarrow{t^*}$ is feasible in $\mathcal{Z}$.

It is clear that m' is reachable in the skeleton of $\mathcal{Z}$ because z' is reachable in $\mathcal{Z}$, i.e., $m' \in R_{S(\mathcal{Z})}$. Then, because of the liveness of $S(\mathcal{Z})$ there is a marking $m'' \in R_{S(\mathcal{Z})}(m')$ with

$$m'' \xrightarrow{t^*} \quad \text{in } S(\mathcal{Z}),$$

which means that there is a firing sequence σ from m' to m'' in $S(\mathcal{Z})$, so

$$m' \xrightarrow{\sigma} m'' \xrightarrow{t^*} \quad \text{in } S(\mathcal{Z}).$$

Let $\sigma = t_1 t_2 \ldots t_n$. We will show by induction on n that the transition sequence σ, possibly modified, is also a firing sequence in $\mathcal{Z}$. After firing of (the possibly modified) σ in $\mathcal{Z}$ the transition t^* can fire in $\mathcal{Z}$.

The idea is the following: For any two consecutive firing transitions t_i and t_{i+1} in σ there are two possibilities in $\mathcal{Z}$. Either the transitions can also fire consecutively in $\mathcal{Z}$ (time can of course, elapse in between the firings), or after the firing of t_i in $\mathcal{Z}$ there are enabled transitions other than t_{i+1} that are forced to fire by their time restrictions before t_{i+1} becomes ready to fire. This is the case if an enabled transition reaches its latest firing time before t_{i+1} reaches its earliest firing time. Such transitions prevent the aging of t_{i+1}. However we will show that there are only finitely many such transitions and that their firing does not disable t_{i+1}.

---

[8]In this book the notion of EFC net is defined for more general nets than usual, cf. definition 2.19.

Basis: $n = 0$, i.e., $m' = m''$ and thus

$$m' = m'' \xrightarrow{t^*} \text{ in } S(\mathcal{Z}). \tag{48}$$

It holds for $z' =: z''$ that $z'' \in RS_{\mathcal{Z}}(z')$. The state z'' was chosen such that t^* is enabled in z''. If it is old enough to fire, i.e., $h''(t^*) \geq eft(t^*)$ then we have proved the claim for the basis, but if t^* has not yet reached its eft in z'', i.e.,

$$h''(t^*) < eft(t^*), \tag{49}$$

then starting in z'' time has to pass in $\mathcal{Z}$ in order for t^* to grow old and reach its $eft(t^*)$. There might be other enabled transitions than t^* that must fire before t^* reaches its eft. We will show that firing such transitions does not affect the enabling of t^* and that after finitely many state changes (by firing or by elapsing of time) a state in which t^* is ready to fire will be reached.

Let

$$\tau_1 := \max \left\{ \tau \mid z'' \xrightarrow{\tau} \text{ feasible in } \mathcal{Z} \right\}$$

Clearly, this can be rewritten as

$$\tau_1 = \min \left\{ lft(t) - h''(t) \mid t \in T \,\wedge\, t^- \leq m'' \right\}$$

and thus, τ_1 can be directly computed.

Assuming that $z'' \xrightarrow{\tau_1} z_1^{(1)}$ in $\mathcal{Z}$ with $z_1^{(1)} = (m_1^{(1)}, h_1^{(1)})$ it holds that

$$m_1^{(1)} = m'' \tag{50}$$

and therefore t^* is also enabled in $z_1^{(1)}$, and

$$h_1^{(1)}(t^*) = h''(t^*) + \tau_1.$$

Case 1: $\tau_1 \geq eft(t^*) - h''(t^*)$.
        Then

$$\begin{aligned} h_1^{(1)}(t^*) &\geq h''(t^*) + eft(t^*) - h''(t^*) \\ &= eft(t^*). \end{aligned}$$

Hence, t^* is ready to fire in $z_1^{(1)}$ in $\mathcal{Z}$, i.e.,

$$z_1^{(1)} \xrightarrow{t^*} \quad \text{is feasible in } \mathcal{Z}.$$

Thus, t^* can fire in this case.

Case 2: $\tau_1 < eft(t^*) - h''(t^*)$.

Then, it holds that:

$$
\begin{aligned}
h_1^{(1)}(t^*) \;&<\; h''(t^*) + eft(t^*) - h''(t^*) \\
&=\; eft(t^*).
\end{aligned} \tag{51}
$$

Let $M_1^{(1)} := \{t \mid t \in T \wedge t^- \leq m_1^{(1)} \wedge h_1^{(1)}(t) = lft(t)\}$, i.e., $M_1^{(1)}$ is the set of all enabled transitions in $z_1^{(1)}$ which have reached their lft in $z_1^{(1)}$ and therefore have to fire.

We now consider the common pre-places of all transitions in $M_1^{(1)}$ and the transition t^*, i.e., we consider the set of places ${}^{\bullet}M_1^{(1)} \cap {}^{\bullet}t^*$.

Case 2.1: ${}^{\bullet}M_1^{(1)} \cap {}^{\bullet}t^* = \emptyset$.

Let $\hat{t}_1^{(1)} \in M_1^{(1)}$. Then, $\hat{t}_1^{(1)}$ is enabled in $z_1^{(1)}$ and has reached its lft in $z_1^{(1)}$. Hence, $\hat{t}_1^{(1)}$ can fire in $z_1^{(1)}$. Let us assume then, that

$$z_1^{(1)} \xrightarrow{\hat{t}_1^{(1)}} z_2^{(1)}.$$

Then because ${}^{\bullet}M_1^{(1)} \cap {}^{\bullet}t^* = \emptyset$, the number of tokens on all pre-places of t^* does not decrease, i.e.,

$$\forall p \big(p \in {}^{\bullet}t^* \longrightarrow m_1^{(1)}(p) \leq m_2^{(1)}(p)\big), \quad \text{so because of (50)}$$
$$\forall p \big(p \in {}^{\bullet}t^* \longrightarrow m_1''(p) \leq m_2^{(1)}(p)\big).$$

Therefore, it holds that

$$t^{*-} \leq m_2^{(1)},$$

i.e., t^* is enabled in $z_2^{(1)}$. For the t-marking $h_2^{(1)}$ it then holds that:

$$h_2^{(1)}(t^*) = h_1^{(1)}(t^*),$$

because the firing of a transition does not take time. Thus, t^* is not ready to fire in $z_2^{(1)}$.

We now again consider the enabled transitions that have reached their lft in the state $z_2^{(1)}$ and the set $M_2^{(1)}$ of all such transitions.

$$M_2^{(1)} := \{t \mid t^- \le m_2^{(1)} \wedge h_2^{(1)}(t) = lft(t)\}.$$

It is possible that there are other transitions enabled in $z_2^{(1)}$ than in $z_1^{(1)}$ but, by condition (d), the newly enabled transitions have not yet reached their lft in $z_2^{(1)}$. Hence, it holds that

$$M_1^{(1)} \supset M_2^{(1)} \quad \text{because} \quad \hat{t}_1^{(1)} \in M_1^{(1)} \text{ and } \hat{t}_1^{(1)} \notin M_2^{(1)}.$$

Therefore there must be another transition $\hat{t}_2^{(1)}$ from $M_2^{(1)}$ which can fire.

Assume that we have already defined $M_k^{(1)}$ with $t_k^{(1)} \in M_k^{(1)}$ and

$$z'' \xrightarrow{\tau_1} z_1^{(1)} \xrightarrow{\hat{t}_1^{(1)}} z_2^{(1)} \xrightarrow{\hat{t}_2^{(1)}} z_3^{(1)} \xrightarrow{\hat{t}_3^{(1)}} \ldots z_k^{(1)} \xrightarrow{\hat{t}_k^{(1)}} z_{k+1}^{(1)}.$$

Then we define $M_{k+1}^{(1)}$ as follows:

$$M_{k+1}^{(1)} := \{t \mid t^- \le m_{k+1}^{(1)} \wedge h_{k+1}^{(1)}(t) = lft(t)\}.$$

It is easy to see that the following sequence of inclusions holds:

$$M_1^{(1)} \supset M_2^{(1)} \supset \ldots \supset M_k^{(1)} \supset M_{k+1}^{(1)} \supset \ldots. \qquad (52)$$

This sequence is obviously finite because $T \supseteq M_1^{(1)}$ and T is finite, i.e., there is a $k_1 \in \mathbb{N}$, $k_1 \le |(\{t \mid t^- \le m''\})| - 1$ and $M_{k_1-1}^{(1)} \ne \emptyset$, but $M_{k_1}^{(1)} = \emptyset$. Let $\hat{\sigma}_1 := \hat{t}_1^{(1)} \hat{t}_2^{(1)} \ldots \hat{t}_{k_1-1}^{(1)}$. So, it holds that

$$z_1^{(1)} \xrightarrow{\hat{\sigma}_1} z_{k_1}^{(1)}.$$

It is evident that there is no transition which has reached its lft in $z_{k_1}^{(1)}$. Hence, from the state $z_{k_1}^{(1)}$ state change through elapsing time is possible, say $z_{k_1}^{(1)} \xrightarrow{\tau_2} z_1^{(2)}$. We repeat this approach and obtain the following feasible run in $\mathcal{Z}$:

$$z'' \xrightarrow{\tau_1} z_1^{(1)} \xrightarrow{\hat{\sigma}_1} z_{k_1}^{(1)} \xrightarrow{\tau_2} z_1^{(2)} \xrightarrow{\hat{\sigma}_2} z_{k_2}^{(2)} \xrightarrow{\tau_3} \ldots \xrightarrow{\hat{\sigma}_r} z_{k_r}^{(r)} \quad (53)$$

If $\tau_1 + \ldots + \tau_r \geq eft(t^*) - h''(t^*)$ then t^* is ready to fire in the state $z_{k_r}^{(r)}$ and thus the assertion is proved for the basis in this case .

Thus, we now assume that $\tau_1 + \ldots + \tau_r < eft(t^*) - h''(t^*)$ and define

$$\tau_{r+1} := \max\{\tau \mid z_{k_r}^{(r)} \xrightarrow{\tau} \text{ feasible in } \mathcal{Z} \}.$$

Similarly as for τ_1 it also holds that

$$\tau_{r+1} = \min\{ lft(t) - h_{k_r}^{(r)}(t) \mid t^- \leq m_{k_r}^{(r)} \}. \quad (54)$$

Thereby we obtain the sequence

$$\tau_1, \tau_2, \ldots, \tau_k, \ldots \quad (55)$$

such that for every natural number k it holds that

$$\sum_{i=1}^{k} \tau_i < eft(t^*) - h''(t^*). \quad (56)$$

The sequence (55) with the property (56) is finite, because:

- Each τ_i is the difference of the lft of a transition enabled in the state $z_{k_{i-1}}^{(i-1)}$ and the time showed by its clock in this state, i.e., $\tau_i = lft(t) - h_{k_{i-1}}^{(i-1)}(t)$ for some $t \in T$.
- Some of the times τ_i might be smaller than 1. This depends on when (in which state) a transition was newly enabled which determines the value τ_i, cf. also (54). Any τ_i which is determined by a transition which was already enabled in z'' can be smaller than 1. There are finitely many of these transitions and therefore there are only finitely many such τ_i.
- Because of condition (d) and because the bounds of all time intervals are natural numbers, all transitions which are newly enabled in a state reached after z'' and which have reached their lft are older than 1.

Now, we recursively define a sequence of indices:

Basis: Let i_1 be the smallest index such that

$$\tau_{i_1} = lft(\tilde{t}_1) - h^{i_1-1}_{k_{i_1}-1}(\tilde{t}_1)$$

for a transition $\tilde{t}_1 \in T$, not enabled in z''. Hence, it holds that:

$$\sum_{j=1}^{i_1} \tau_j \geq 1.$$

Step: Let i_λ be defined already. Then, we let $i_{\lambda+1}$ be the smallest index with

- $i_\lambda < i_{\lambda+1}$ and
- $\tau_{i_{\lambda+1}} = lft(\tilde{t}_{\lambda+1}) - h^{(i_{\lambda+1}-1)}_{k_{i_{\lambda+1}}-1}(\tilde{t}_{\lambda+1})$, where $\tilde{t}_{\lambda+1}$ was last enabled in a state reached in (53) after the state $z_1^{(i_\lambda)}$.

Hence, it holds that:

$$\sum_{j=i_\lambda+1}^{i_{\lambda+1}} \tau_j \geq 1.$$

After at most $\lceil eft(t^*) - h''(t^*) \rceil$ such indices the inequality (56) is no longer fulfilled. Consequently, after finitely many state changes a state is reached where t^* is enabled and has achieved its eft. Hence, t^* is ready to fire in that state.

Thus, the assertion has been proved for the basis in this case.

Case 2.2: $^\bullet M_1^{(1)} \cap {}^\bullet t^* \neq \emptyset$.

This means that there is a transition $\hat{t}$ which has reached its lft in $z_1^{(1)}$ and has at least one common pre-place with t^*. We will show that this is not possible.

Because $\hat{t} \in M_1^{(1)}$, it holds that:

$$lft(\hat{t}) = h_1^{(1)}(\hat{t}). \tag{57}$$

and

$$^\bullet\hat{t} \cap {}^\bullet t^* \neq \emptyset. \tag{58}$$

Together with condition (*a*) this yields that

$$\bullet \hat{t} = \bullet t^*.$$

Furthermore, because of condition (*b*) it holds that $\hat{t}$ is enabled if and only if t^* is enabled. Therefore it follows that

$$h_1^{(1)}(t^*) = h_1^{(1)}(\hat{t}). \tag{59}$$

Now, condition (*c*) and (57) lead to

$$eft(t^*) \leq Min(p) \leq Max(p) \leq lft(\hat{t})$$

for every $p \in \bullet t^* = \bullet \hat{t}$. Together with (59) this leads to

$$h_1^{(1)}(t^*) \geq eft(t^*),$$

which is a contradiction to (51) (which holds in Case 2).

Thus, we have proved for the induction basis that t^* can fire.

Step: As *induction hypothesis* we assume that the assertion is true for $\ell(\sigma) \leq n$, i.e., if in $S(\mathcal{Z})$ the transition t^* can fire after the firing of a transition sequence σ started in m', with $\ell(\sigma)$ at most n, then the transition t^* can also fire in $\mathcal{Z}$ after the firing of a sequence $mod(\sigma)$, derived from σ started in z'.

The *induction claim* is that if

$$m' \xrightarrow{\sigma} m'' \xrightarrow{t^*} \quad \text{with} \quad \sigma = t_1 \ldots t_n t_{n+1}$$

is feasible in $S(\mathcal{Z})$, then there is a firing sequence $mod(\sigma)$ in $\mathcal{Z}$, derived from σ, such that after firing $mod(\sigma)$ the transition t^* is ready to fire, i.e.,

$$z' \xrightarrow{mod(\sigma)} \xrightarrow{t^*} \quad \text{is feasible in } \mathcal{Z}.$$

We denote the sequence $t_2 \ldots t_n t_{n+1}$ by $\tilde{\sigma}$ and the sequence $t_2 \ldots t_n t_{n+1} t^*$ by $\tilde{\tilde{\sigma}}$ for the rest of this proof.

Thus, now it suffices to prove that

$$z' \xrightarrow{\ mod(t_1)\ } \tilde{z} = (\tilde{m}, \tilde{h}) \text{ can fire in } \mathcal{Z}$$

where

$$\tilde{m} \xrightarrow{\ mod(\tilde{\sigma})\, t^*\ } \text{ in } S(\mathcal{Z}) \text{ with } \ell(mod(\tilde{\sigma})) \leq n. \tag{60}$$

The induction claim will consequently follow by the induction hypothesis and thus the theorem will be proved.

Let us now consider the state z'. If $h'(t_1) \geq eft(t_1)$, then t_1 can fire in z' in $\mathcal{Z}$, i.e., $mod(t_1) = t_1$ and $mod(\tilde{\sigma}) = \tilde{\sigma}$. Then, (60) is obviously true and therefore in this case the theorem holds. Thus, we consider the case that

$$h'(t_1) < eft(t_1). \tag{61}$$

Let

$$\theta_1 := \max\{\, \theta \mid z' \xrightarrow{\ \theta\ } \text{ feasible in } \mathcal{Z} \,\}.$$

Then, it holds that:

$$\theta_1 = \min\{\, lft(t) - h'(t) \mid t \in T \ \wedge\ t^- \leq m' \,\}$$

and therefore θ_1 can easily be computed.

Let $z' \xrightarrow{\ \theta_1\ } \tilde{z}_1^{(1)}$ in $\mathcal{Z}$ with $\tilde{z}_1^{(1)} = (\tilde{m}_1^{(1)}, \tilde{h}_1^{(1)})$. Obviously, t_1 is also enabled in $\tilde{z}_1^{(1)}$ because $\tilde{m}_1^{(1)} = m'$.

Case 1: $\theta_1 \geq eft(t_1) - h'(t_1)$.
Then, it follows that

$$\tilde{h}_1^{(1)}(t_1) = h'(t_1) + \theta_1 \geq eft(t_1)$$

and hence t_1 is ready to fire in $\tilde{z}_1^{(1)}$. Evidently, (60) is fulfilled with $\tilde{z} := \tilde{z}_1^{(1)}$, $mod(t_1) = t_1$ and $mod(\tilde{\sigma}) = \tilde{\sigma}$. Thereby, the assertion of the theorem is proved for this case.

Case 2: $\theta_1 < eft(t_1) - h'(t_1)$.

We define the set $N_1^{(1)}$ (similar to $M_1^{(1)}$)

$$N_1^{(1)} := \{\, t \mid t \in T \,\wedge\, t^- \leq \tilde{m}_1^{(1)} \,\wedge\, \tilde{h}_1^{(1)}(t) = lft(t)\,\}$$

and consider all common pre-places of transitions in $N_1^{(1)}$ and $\tilde{\tilde{\sigma}}$:

Case 2.1: $\bullet N_1^{(1)} \cap \bullet\tilde{\tilde{\sigma}} \neq \emptyset$.

Let $\tilde{t}_1^{(1)} := t_j$ where $j := \min\{\, r \mid t_r \in \tilde{\tilde{\sigma}} \,\wedge\, \bullet N_1^{(1)} \cap \bullet t_r \neq \emptyset \,\}$, i.e.,

- $\tilde{t}_1^{(1)}$ has achieved its lft in $\tilde{z}_1^{(1)}$,
- $\tilde{t}_1^{(1)} = t^*$ or $\tilde{t}_1^{(1)}$ belongs to the sequence $\tilde{\sigma}$,
- $\tilde{t}_1^{(1)}$ is the transition with the smallest index in $\tilde{\tilde{\sigma}}$, with these properties.

It is clear, that $\bullet t_j \cap \bullet t_1 = \emptyset$, because otherwise t_1 would have already reached its eft in $\tilde{z}_1^{(1)}$ in contradiction to the assumption of Case 2.

Without loss of generality let $\tilde{t}_1^{(1)} \neq t^*$ because otherwise the claim already holds. Let

$$\tilde{z}_1^{(1)} \xrightarrow{\tilde{t}_1^{(1)}} \tilde{z}_2^{(1)} \quad \text{in} \quad \mathcal{Z}.$$

By the choice of j it holds that

$$\bullet\tilde{t}_1^{(1)} \cap \bullet t_l = \emptyset \quad \text{for all} \quad t_l \quad \text{in} \quad \tilde{\tilde{\sigma}} \quad \text{with} \quad 1 < l < j. \qquad (62)$$

Now let t_s be in $\tilde{\tilde{\sigma}}$ with $s > j$.

Case 2.1.1: $\bullet\tilde{t}_1^{(1)} \cap \bullet t_s = \emptyset$,

Because of (62) and the assumption for this case, it holds that when $\tilde{t}_1^{(1)}$ fires then the number of tokens in the pre-places of all transitions in

$$t_1(\tilde{\sigma} \setminus \tilde{t}_1^{(1)}) = t_1 t_2 \ldots t_{j-1} t_{j+1} \ldots t_{n+1} =: mod(\tilde{\sigma})$$

does not decrease. Thus, we have that

$$\tilde{m}_1^{(1)} \xrightarrow{\tilde{t}_1^{(1)}} \tilde{m}_2^{(1)} \xrightarrow{mod(\tilde{\sigma})} \xrightarrow{t^*} \quad \text{is feasible in } S(\mathcal{Z}).$$

Hereby, the number of transitions preventing the aging of t_1 is decreased by 1. Furthermore, it holds for the length of $mod(\tilde{\sigma})$ that $\ell(mod(\tilde{\sigma})) = \ell(\sigma) - 1 = n$. Thus, the induction hypothesis can be applied to $\tilde{z} := \tilde{z}_2^{(1)}$ and $mod(\tilde{\sigma})$. Consequently, the induction claim follows and thereby the assertion of the theorem is proved for this case, as well.

Case 2.1.2: $^\bullet \tilde{t}_1^{(1)} \cap {}^\bullet t_s \neq \emptyset.$
This means that each transition $t_s, s > j$ becomes disabled (at least momentarily) after the firing of $\tilde{t}_1^{(1)} = t_j$. Nevertheless, all these transitions become enabled again at the latest after the firing of t_{j-1}, because it holds that

$$\tilde{m}_2^{(1)} \xrightarrow{t_j t_1 t_2 \dots t_{j-1}} m_j \quad \text{in } S(\mathcal{Z})$$

and

$$\tilde{m}_2^{(1)} \xrightarrow{t_1 t_2 \dots t_{j-1} t_j} m_j \quad \text{in } S(\mathcal{Z}).$$

Therefore it immediately follows for the sequence $mod(\tilde{\sigma}) := t_1(\tilde{\sigma} \setminus \tilde{t}_1^{(1)})$ that

$$\tilde{m}_2^{(1)} \xrightarrow{mod(\tilde{\sigma})} \xrightarrow{t^*} \quad \text{is feasible in } S(\mathcal{Z})$$

and $\ell(mod(\tilde{\sigma})) = n$. Thus, the induction hypothesis can be applied to $\tilde{z} := \tilde{z}_2^{(1)}$ and $mod(\tilde{\sigma})$ and once again the induction claim follows and thereby the claim of the theorem is proved for this case.

Case 2.2: $^\bullet N_1^{(1)} \cap {}^\bullet \tilde{\tilde{\sigma}} = \emptyset.$
Let $\tilde{t}_1^{(1)} \in N_1^{(1)}$. Hence, $\tilde{h}_1^{(1)}(\tilde{t}_1^{(1)}) = lft(\tilde{t}_1^{(1)})$ and $\tilde{t}_1^{(1)}$ is ready to fire in $\tilde{z}_1^{(1)}$. Let

$$\tilde{z}_1^{(1)} \xrightarrow{\tilde{t}_1^{(1)}} \tilde{z}_2^{(1)} \quad \text{be feasible in } \mathcal{Z}.$$

By the assumption for this case the number of tokens in the pre-places of $\tilde{\tilde{\sigma}}$ does not decrease after firing of $\tilde{t}_1^{(1)}$, i.e.,

$$\forall p \left(p \in {}^{\bullet}\tilde{\sigma} \; \longrightarrow \; m'(p) = \tilde{m}_1^{(1)}(p) \leq \tilde{m}_2^{(1)}(p) \right). \tag{63}$$

Moreover, t_1 and $\tilde{t}_1^{(1)}$ have no common pre-places, because otherwise t_1 would have reached its lft in $\tilde{z}_1^{(1)}$ which is a contradiction to the assumption for Case 2. Therefore it holds that

$$\forall p \left(p \in {}^{\bullet}t_1 \; \longrightarrow \; m'(p) = \tilde{m}_1^{(1)}(p) \leq \tilde{m}_2^{(1)}(p) \right). \tag{64}$$

Similarly to the induction basis we define $N_j^{(1)}$ where $\tilde{t}_j^{(1)} \in N_j^{(1)}$ but $\tilde{t}_j^{(1)} \notin N_{j+1}^{(1)}$. Then it holds that

$$z' \xrightarrow{\theta_1} \tilde{z}_1^{(1)} \xrightarrow{\tilde{t}_1^{(1)}} \tilde{z}_2^{(1)} \xrightarrow{\tilde{t}_2^{(1)}} \ldots \xrightarrow{\tilde{t}_{r_1}^{(1)}} \tilde{z}_{r_1}^{(1)}$$

and $N_{r_1}^{(1)} = \emptyset$, i.e., from the state $\tilde{z}_{r_1}^{(1)}$ state change through elapsing time is possible. Moreover, for all j with $1 \leq j \leq r_1$ it always holds that:

$$\forall p \left(p \in ({}^{\bullet}t_1 \cup {}^{\bullet}\tilde{\sigma}) \; \longrightarrow \; m'(p) \leq \tilde{m}_j^{(1)}(p) \leq \tilde{m}_{j+1}^{(1)}(p) \right). \tag{65}$$

We can now define θ_2, the amount of time to elapse:

$$\theta_2 := \max\{\theta \mid \tilde{z}_{r_1}^{(1)} \xrightarrow{\theta} \text{ feasible in } \mathcal{Z} \}.$$

Again, similarly as in the induction basis, we obtain the sequence

$$z' \xrightarrow{\theta_1} \tilde{z}_1^{(1)} \xrightarrow{\tilde{\sigma}_1} \tilde{z}_{r_1}^{(1)} \xrightarrow{\theta_2} \tilde{z}_1^{(2)} \xrightarrow{\tilde{\sigma}_2} \tilde{z}_{r_2}^{(2)} \xrightarrow{\theta_3} \ldots \tag{66}$$

with

$$\tilde{\sigma}_i = \tilde{t}_1^{(i)} \tilde{t}_2^{(i)} \ldots \tilde{t}_{r_i}^{(i)}.$$

In each state $\tilde{z}_1^{(i)}, i = 1, 2, \ldots$ each transition $\tilde{t}_j^{(i)}$ for $j = 1, \ldots, r_i$ has reached its lft and prevents the aging of t_1. After the firing of all the transitions $\tilde{t}_j^{(i)}, j = 1, \ldots, r_i$ the elapsing of time θ_{i+1} is possible and t_1 can age.

Similarly as in the induction basis, a run such as in (66) is always finite, i.e., there is a natural number s such that

$$z' \xrightarrow{\theta_1} \tilde{z}_1^{(1)} \xrightarrow{\tilde{\sigma}_1} \tilde{z}_{r_1+1}^{(1)} \xrightarrow{\theta_2} \tilde{z}_1^{(2)} \xrightarrow{\tilde{\sigma}_2} \tilde{z}_{r_2}^{(2)} \xrightarrow{\theta_3} \dots$$
$$\tilde{z}_{r_s}^{(s)} \xrightarrow{\theta_s} \tilde{z}_1^{(s+1)} \xrightarrow{t_1} \tag{67}$$

i.e., (67) is a feasible run in $\mathcal{Z}$.

If during the firing of (67) a transition $\tilde{t}_j^{(i)}$ belonging to $\tilde{\tilde{\sigma}}$ fires, then we need to consider the run (67) only up to the state change

$$\tilde{z}_j^{(i)} \xrightarrow{\tilde{t}_j^{(i)}} \tilde{z}_{j+1}^{(i)}.$$

This is because the induction hypothesis can be applied to

$$\tilde{z} := \tilde{z}_{j+1}^{(i)} \text{ and } mod(\tilde{\sigma}) := t_1 \left(\tilde{\sigma} \setminus t_j \right) \text{ with } t_j = \tilde{t}_j^{(i)},$$

analogously to Case 2.1. Therefore

$$mod(t_1) = \tilde{\sigma}_1 \dots \tilde{\sigma}_{i-1} \tilde{t}_1^{(i)} \dots \tilde{t}_j^{(i)}.$$

If during the firing of (67) none of the transitions $\tilde{t}_j^{(i)}$ for $i = 1, \dots, s$, $j = 1, \dots, r_i$, belonging to $\tilde{\tilde{\sigma}}$ fires, then the transition t_1 can fire in the state $\tilde{z}_1^{(s+1)}$. Thus, the induction hypothesis can now be applied to $\tilde{z} := \tilde{z}_1^{(s+1)}$ and $mod(\tilde{\sigma}) = \tilde{\sigma}$. Therefore, $mod(t_1) = \tilde{\sigma}_1 \dots \tilde{\sigma}_s$ and this concludes the inductive step and the proof of the theorem.   □

Remark 3.62 *In a Time Petri net for which the conditions of Theorem 3.61 hold, the set of all reachable p-markings in $\mathcal{Z}$ can be a proper subset of the set of markings reachable in the skeleton of $\mathcal{Z}$, i.e., $R_{\mathcal{Z}} \subset R_{S(\mathcal{Z})}$.*

Proof:

For an arbitrary Time Petri net $\mathcal{Z}$ it holds that $R_{\mathcal{Z}} \subseteq R_{S(\mathcal{Z})}$. We now consider the Time Petri net $\mathcal{Z}_7$ in Fig. 3.21 with $R_{\mathcal{Z}_7} \neq R_{S(\mathcal{Z}_7)}$.

Obviously, $\mathcal{Z}_7$ is a Time Petri net satisfying the conditions of Theorem 3.61. It is easy to see that $(0,0,1,1,0) \in R_{S(\mathcal{Z})}$ but $(0,0,1,1,0) \notin R_{\mathcal{Z}}$.   □

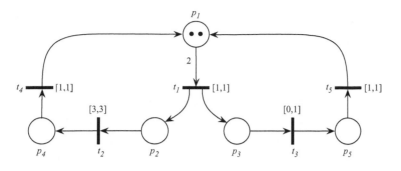

Figure 3.21: The Time Petri net $\mathcal{Z}_7$

In Figs. 3.22, 3.23, 3.24 and 3.25 we now consider four Time Petri nets $\mathcal{Z}_{(i)}$ of which each contradicts the assumption that Theorem 3.61 might still hold if we omit condition $i \in \{a, b, c, d\}$ respectively.

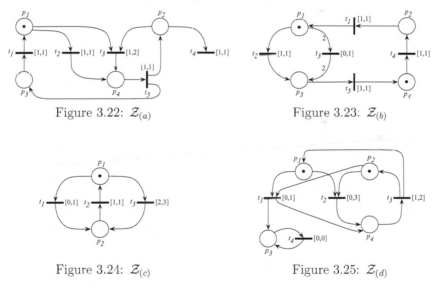

Figure 3.22: $\mathcal{Z}_{(a)}$ Figure 3.23: $\mathcal{Z}_{(b)}$

Figure 3.24: $\mathcal{Z}_{(c)}$ Figure 3.25: $\mathcal{Z}_{(d)}$

It is easy to see that the skeletons of the Time Petri nets $\mathcal{Z}_{(a)}$, $\mathcal{Z}_{(b)}$, $\mathcal{Z}_{(c)}$ and $\mathcal{Z}_{(d)}$ are live, but the nets themselves are not. Moreover, the net $S(\mathcal{Z}_{(a)})$ in addition to not being an EFC net is also not an AC net. Actually, it can be proved that Theorem 3.61 still holds if we require $S(\mathcal{Z})$ to be an AC net instead of an EFC net (cf. [BPZ10] and in [Bac11] or try proving this as an exercise). The proof of Theorem 3.61 however cannot be generalized to AC nets because we cannot draw conclusion (59) for AC nets.

We presume that using statical properties it is hardly possible to define a larger class of Time Petri nets for which Theorem 3.61 holds than the class of AC nets with properties $(b), (c)$ and (d) from the theorem. We can however introduce a dynamic extension. To this end, we define *Behaviorally Free Choice nets (BFC nets)* which were first introduced by E. Best in [Bes87].

Definition 3.63 (BFC net) *A Petri net $\mathcal{N} = (P, T, F, V, m_0)$ is a BFC net if for all transitions t_1 and t_2 from T with at least one common pre-place and for every reachable marking m in $\mathcal{N}$ it holds that: t_1 is enabled in m if and only if t_2 is enabled in m.*

It is easy to see that every EFC net is also a BFC net. The following example shows that the converse does not hold.

Example 3.64 *In the Petri net $S(\mathcal{Z}_8)$ it holds that: $^\bullet t_1 = \{p_1, p_2\}$, $^\bullet t_2 = \{p_2, p_3\}$, i.e., $S(\mathcal{Z}_8)$ is not a EFC net.*

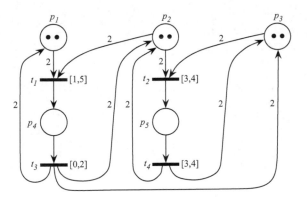

Figure 3.26: The Time Petri net $\mathcal{Z}_8$

Only the transitions t_1 und t_2 have a common pre-place in $S(\mathcal{Z}_8)$, namely, $^\bullet t_1 \cap {}^\bullet t_2 = \{p_2\}$ and in any reachable marking either both transitions are enabled or both are disabled.

So the set of all BFC nets is obviously a proper superset of the set of all EFC nets. But the EFC property is a statical property and therefore decidable whereas the BFC property is a dynamic property. It is in general not decidable for an unbounded Petri net whether it is a BFC net or not.

Theorem 3.65 *Let $\mathcal{Z} = (P, T, F, V, m_0, I)$ be a Time Petri net such that*

(a) $S(\mathcal{Z})$ is a BFC-net,

(b) $S(\mathcal{Z})$ is homogeneous,

(c) for each place $p \in P$ it holds that: $\mathcal{M}in(p) \leq \mathcal{M}ax(p)$,

(d) for each transition $t \in T$ it holds that: $lft(t) > 0$,

(e) $\forall t_1 \forall t_2 \, (\, {}^\bullet t_1 \cap {}^\bullet t_2 \neq \emptyset \longrightarrow \forall t_3 \, \forall i \, \forall j \, (\, (\, i, j \in \{1, 2\} \wedge i \neq j \wedge {}^\bullet t_i \cap {}^\bullet t_3 \neq \emptyset) \longrightarrow {}^\bullet t_j \cap {}^\bullet t_3 \neq \emptyset)).$

Then if $S(\mathcal{Z})$ is live, $\mathcal{Z}$ is live, too.

Remark 3.66 *Clearly, the Time Petri net $\mathcal{Z}_8$ fulfills all conditions of Theorem 3.65. Therefore and by 3.64, the set of nets defined in Theorem 3.65 is a proper superset of the set of nets defined in Theorem 3.61.*

Proof of Theorem 3.65:

We prove the theorem in two steps:

In Step 1 we construct a new Petri net $\mathcal{N}$ derived from $S(\mathcal{Z})$. The new net will be an EFC-net which is live if and only if $S(\mathcal{Z})$ is live. Afterwards, a second Time Petri net $\mathcal{Z}'$ with $S(\mathcal{Z}') = \mathcal{N}$ is considered. The net $\mathcal{Z}'$ will comply with the conditions of Theorem 3.61. Hence, $\mathcal{Z}'$ will be live if $\mathcal{N}$ is live, i.e., if $S(\mathcal{Z})$ is live then $\mathcal{N}$ and $\mathcal{Z}'$ are live, too.

In Step 2 the proof will be concluded by showing that $\mathcal{Z}$ is live if and only if $\mathcal{Z}'$ is live.

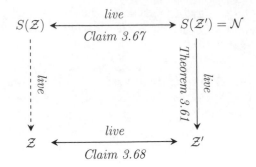

Figure 3.27: Structure of the proof of Theorem 3.65

Step 1:

Let $\mathcal{Z} = (P, T, F, V, m_0, I)$ be a Time Petri net satisfying conditions $(a), (b),$ $(c), (d)$ and (e). Its skeleton $S(\mathcal{Z})$ is the Petri net (P, T, F, V, m_0).

The new Petri net $\mathcal{N} := (P, T, F', V', m_0)$ derived from $S(\mathcal{Z})$ is defined by

$$F' := F \cup \{(p, t), (t, p) \mid \exists p \, \exists q \, (p, q \in P \wedge p^\bullet \cap q^\bullet \neq \emptyset \wedge t \in q^\bullet \wedge t \notin p^\bullet)\}$$

and

$$V'(f) := \begin{cases} V(f) & \text{if } f \in F \\ V(\tilde{f}) & \text{if } f \notin F \wedge \big(f = (p, t) \vee f = (t, p)\big) \wedge \\ & \qquad \tilde{f} = (p, \tilde{t}) \text{ for some } \tilde{t} \in T. \end{cases}$$

The Petri net $\mathcal{N}$ is obviously an EFC net. Furthermore, it follows from Claim 3.67, given below, that the nets $S(\mathcal{Z})$ and $\mathcal{N}$ are either both live or both not live.

To illustrate how $\mathcal{N}$ is derived from $S(\mathcal{Z})$ we give an example. The Petri net $\mathcal{N}$, derived from the skeleton $S(\mathcal{Z}_8)$ is the skeleton of the Time Petri net $\mathcal{Z}_8'$ shown in Figure 3.28.

Claim 3.67 *For the Petri nets $S(\mathcal{Z})$ and $\mathcal{N}$, and for every transition sequence $\sigma \in T^*$ it holds that*

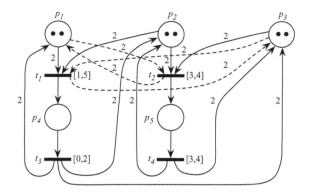

Figure 3.28: The Time Petri net $\mathcal{Z}'_8$ derived from $\mathcal{Z}_8$ by adding the dashed arcs as defined in the construction of $\mathcal{N}$ in Step 1

$$m_0 \xrightarrow{\sigma} m' \text{ in } S(\mathcal{Z}) \text{ if and only if } m_0 \xrightarrow{\sigma} m' \text{ in } \mathcal{N}.$$

The proof can easily be done by induction on the length of σ.   ✓

We now denote the Time Petri net (P, T, F', V', m_0, I) by $\mathcal{Z}'$.

Step 2:

The following claim states that the Time Petri nets $\mathcal{Z}$ and $\mathcal{Z}'$ are either both live or both not live.

Claim 3.68 *For the Time Petri nets $\mathcal{Z}$ and $\mathcal{Z}'$, and for any transition sequence $\sigma \in T^*$, it holds that:*

$$z_0 \xrightarrow{\sigma} z' \text{ in } \mathcal{Z} \text{ if and only if } z_0 \xrightarrow{\sigma} z' \text{ in } \mathcal{Z}'.$$

Proof:

($\Longrightarrow$)

Let $\sigma = t_1 \ldots t_n$ such that $z_0 \xrightarrow{\sigma} z'$ in $\mathcal{Z}$. We have to show that $z_0 \xrightarrow{\sigma} z'$ is feasible in $\mathcal{Z}'$, too. We prove this claim by induction on $\ell(\sigma)$:

Basis: $\ell(\sigma) = 0$, i.e., $\sigma = \varepsilon$

The claim follows immediately because by construction of $\mathcal{Z}'$ it holds that $m_0 \in R_{\mathcal{Z}}$ and $m_0 \in R_{\mathcal{Z}'}$, and therefore also $z_0 = z_0'$.

Step: *Induction hypothesis:*   The claim holds for any transition sequence $\sigma \in T^*$ of length $\ell(\sigma) = n$, so

$$\text{if } z_0 \xrightarrow{\ t_1...t_n\ } z' \text{ is feasible in } \mathcal{Z},$$

$$\text{then } z_0 \xrightarrow{\ t_1...t_n\ } z' \text{ is feasible in } \mathcal{Z}', \text{ too.} \tag{68}$$

Induction claim: The claim holds for all $\sigma \in T^*$ of length $\ell(\sigma) = n+1$.

$$\text{If } z_0 \xrightarrow{\ t_1...t_n t_{n+1}\ } z' \text{ is feasible in } \mathcal{Z}, \tag{69}$$

$$\text{then } z_0 \xrightarrow{\ t_1...t_n t_{n+1}\ } z' \text{ is feasible in } \mathcal{Z}', \text{ too.} \tag{70}$$

Let us assume that (69) holds. We consider the sequence $t_1 \ldots t_n$ in $\mathcal{Z}$. It holds that:

$$\text{If } z_0 \xrightarrow{\ t_1...t_n\ } z_n \text{ is feasible in } \mathcal{Z}, \tag{71}$$

and by the induction hypothesis for the transition sequence $t_1 \ldots t_n$ it also holds that

$$z_0 \xrightarrow{\ t_1...t_n\ } z_n \text{ is feasible in } \mathcal{Z}'. \tag{72}$$

Because of (69), (71) and (72) it follows that t_{n+1} is enabled in the state z_n (in $\mathcal{Z}'$). Because of (69) and (71) it also follows that there is a non-negative rational number τ_{n+1} with

$$z_n \xrightarrow{\ \tau_{n+1}\ } z_n' \xrightarrow{\ t_{n+1}\ } z_{n+1} \text{ in } \mathcal{Z}. \tag{73}$$

Hence, it holds that:

$$h_n'(t) = h_n(t) + \tau_{n+1} \leq \min\{\, lft(t) \mid t \in T \ \wedge \ t^- \leq m_n \,\} \text{ in } \mathcal{Z}. \tag{74}$$

Therefore and by (73), it follows that in $\mathcal{Z}$:

$$eft(t_{n+1}) \leq h_n'(t_{n+1}). \tag{75}$$

Because of conditions (a) and (b) and the construction of the Time Petri net $\mathcal{Z}'$, the inequality (74) is also true in $\mathcal{Z}'$. Hence, the inequality (75) also holds in $\mathcal{Z}'$, i.e., the state z'_n is also reachable in $\mathcal{Z}'$. Moreover, the transition t_{n+1} is also ready to fire in $\mathcal{Z}'$

We have therefore proved the first direction of the claim. The second direction follows similarly. $\checkmark$

It follows from Claim 3.68 and Definition 3.57 that $\mathcal{Z}$ is live if and only if $\mathcal{Z}'$ is live and together with Claim 3.67 this completes the proof of Theorem 3.65. $\square$

In addition to the conditions of Theorem 3.61, Theorem 3.65 requires condition (e) for the skeleton. This is necessary to ensure that two enabled transitions in the Time Petri net with a common pre-place also become ready to fire. We can omit condition (e) if we define *Behaviourly Free Choice nets* more strictly in Definition 3.63 replacing "enabled" by "newly enabled" which is defined as follows:

Definition 3.69 (newly enabled) *Let $\mathcal{N}$ be a Petri net and $t, \hat{t}$ transitions in $\mathcal{N}$. Let m_1 and m_2 be markings in $\mathcal{N}$ with $m_1 \xrightarrow{\hat{t}} m_2$. The transition t is called newly enabled in m_2 if it is true that $t^- \leq m_2$ and*

1. $t^- \not\leq m_1$ or

2. $t^- \leq m_1$ and $t^- \not\leq m_1 - \hat{t}^-$.

3.7.3 T-invariants

We want to consider the computation of T-invariants in Time Petri nets. A *feasible T-invariant* in a Time Petri net is the Parikh vector of a firing sequence σ, such that after firing some run of σ the net is in the same state as before the firing. The reason for this is pragmatic: Models of biochemical networks can be represented as Time Petri nets. The existence of a T-invariant in a Time Petri net ensures the existence of a so called *steady state*

in the real system, i.e., the biochemical system can be stabilized in a cycle
of some processes (cf. [PZHK05]).

Lemma 3.70 *Let $\mathcal{Z}$ be a Time Petri net and σ a firing sequence in $S(\mathcal{Z})$
with the Parikh vector $\pi_\sigma = x$ such that x is a T-invariant in $S(\mathcal{Z})$. If the
system of inequalities B_σ is solvable in $\mathbb{R}$, then x is also a feasible T-invariant
in $\mathcal{Z}$.*

This lemma follows immediately from *Problem 1* in the next section. The
essential point is the fact that feasible T-invariants can also be computed for
unbounded Time Petri nets or without knowing the state space of the net (cf.
solution of Problem 1). This is particularly important in the area of metabolic
networks, because most of the Time Petri net models of biochemical networks
are unbounded.

3.8   Quantitative Evaluation

Quantitative evaluation is a fundamental part of the analysis of time-dependent
systems. It provides information about the behavior of the system over time.
Moreover, quantitative evaluation can be used to verify the model of a considered system. In [PZ11] we show how to use quantitative evaluation of
Time Petri nets in order to verify that a net is a suitable model for a specific
biochemical network. Time is of course not the only means of verification
but it is an important one.

In this section we will consider problems such as computing minimum and
maximum lengths of time of feasible runs, minimum and maximum distances
of time from one state or place-marking to another and deciding whether a
feasible run of a given length of time exists.

We first introduce some notions that are fundamental in quantitative evaluation:

Definition 3.71 (length of a run) *Let $\mathcal{Z} = (P, T, F, V, m_0, I)$ be a Time
Petri net and $\sigma(\tau)$ a feasible run of the firing sequence σ in $\mathcal{Z}$. The length
of time $\ell(\sigma(\tau))$ of $\sigma(\tau)$ is the sum of all times elapsing over the course of
the firing of $\sigma(\tau)$, formally:*

$$\ell\big(\sigma(\tau)\big) := \sum_{i=0}^{n} \tau_i, \quad \text{for } n = \ell(\sigma) \ \text{and} \ \tau = \tau_0 \tau_1 \ldots \tau_n.$$

Definition 3.72 (minimum run) *Let $\mathcal{Z}$ be a Time Petri net and σ a firing sequence in $\mathcal{Z}$. The feasible run $\sigma(\tau^*)$ of σ has minimum length of time (short: $\sigma(\tau^*)$ is a minimum run of σ), if there is no feasible run of σ with length of time shorter than $\ell(\sigma(\tau^*))$.*

For a minimum run $\sigma(\tau^*)$ of σ it obviously holds that

$$\ell\big(\sigma(\tau^*)\big) = \min_{\tau}\{\,\ell\big(\sigma(\tau)\big) \mid \sigma(\tau) \text{ is a feasible run of } \sigma \text{ in } \mathcal{Z}\,\}.$$

The notion of *maximum run* can be defined similarly, if the set of all lengths of feasible runs of σ has an upper bound. Clearly, even though a run is a finite sequence consisting of transitions and elapses of time the set of all feasible runs of σ might be infinite and therefore the set of their lengths can be infinite and possibly unbounded. In such a case no maximum run of σ exists.

Definition 3.73 (maximum run) *Let $\mathcal{Z}$ be a Time Petri net and σ a firing sequence in $\mathcal{Z}$. The feasible run $\sigma(\tau^*)$ of σ has maximum length of time (short: $\sigma(\tau^*)$ is a maximum run of σ), if the set*

$$\{\,\ell\big(\sigma(\tau)\big) \mid \sigma(\tau) \text{ is a feasible run of } \sigma \text{ in } \mathcal{Z}\,\}$$

has an upper bound and

$$\ell\big(\sigma(\tau^*)\big) = \sup_{\tau}\{\,\ell\big(\sigma(\tau)\big) \mid \sigma(\tau) \text{ is a feasible run of } \sigma \text{ in } \mathcal{Z}\,\}.$$

Obviously for an arbitrary firing sequence, the minimum as well as the maximum run, if they exist, are in general not uniquely determined.

Definition 3.74 (minimum distance of time between states) *Let $\mathcal{Z}$ be a Time Petri net and z_1 and z_2 reachable states in $\mathcal{Z}$ with $z_2 \in RS_{\mathcal{Z}}(z_1)$. The minimum distance of time $d_{\min}(z_1, z_2)$ from z_1 to z_2 is the minimum length of time of any minimum run from z_1 to z_2, formally:*

$$d_{\min}(z_1, z_2) := \min_{\sigma} \{ \ell(\sigma(\tau)) \ \mid \ \sigma \ \textit{is a transition sequence in} \ \mathcal{Z},$$

$$\sigma(\tau) \ \textit{is a minimum run of} \ \sigma \ \textit{and}$$

$$z_1 \xrightarrow{\sigma(\tau)} z_2 \ \textit{in} \ \mathcal{Z} \}$$

The notion of *maximum distance of time between states* is defined differently because for a run starting in z_1 which can get caught in a cycle[9] and never reach z_2 we want to define the maximum distance of time to be infinite.

Definition 3.75 (maximum distance of time between states) *Let $\mathcal{Z}$ be a Time Petri net and z_1 and z_2 reachable states in $\mathcal{Z}$ with $z_2 \in RS_{\mathcal{Z}}(z_1)$. The maximum distance of time $d_{\max}(z_1, z_2)$ from z_1 to z_2 is defined as follows:*

$$d_{\max}(z_1, z_2) := \begin{cases} \infty & \textit{if starting in } z_1 \textit{ a cycle or a dead state} \\ & \quad \textit{can be reached before reaching } z_2 \\ \max_{\substack{\sigma, \\ z_1 \xrightarrow{\sigma(\tau)} z_2}} \sum_{i=0}^{\ell(\sigma)} \tau_i & \textit{otherwise.} \end{cases}$$

The symbol ∞ is used here in order to emphasize that there is a run starting in z_1 and never reaching z_2 because of reaching either a cycle or a dead state. It is also possible that a cycle is reached when all runs from z_1 to z_2 consists only of edges weighted with zero. In that case the length of every such run is also zero but $d_{\max}(z_1, z_2) = \infty$.

Before we continue it should be noted that the maximum distance of time between states is not equivalent to the *longest-path-length* between vertices, which is the maximum length of any simple path in a given graph, cf. [CLRS01]. Actually there even exists a *linear time algorithm* computing the maximum distance of time between states, whereas the longest-path-length problem is NP-hard.

Let m_1 and m_2 be p-markings with $m_1 \in R_{\mathcal{Z}}$ and $m_2 \in R_{\mathcal{Z}}(m_1)$. We will use the following two sets for the next definitions:

$$M_1 := \{ z \mid z \in RS_{\mathcal{Z}} \ \wedge \ \exists h(z = (m_1, h)) \}.$$
$$M_2 := \{ z \mid \exists z^* \left(z^* \in M_1 \ \wedge \ z \in RS_{\mathcal{Z}}(z^*) \ \wedge \ \exists h(z = (m_2, h)) \right) \}.$$

---

[9]A cycle is a finite sequence of successive states in a run such that the first and the last state are the same.

Definition 3.76 (minimum distance of time between p-markings) *Let $\mathcal{Z}$ be a Time Petri net and m_1 and m_2 reachable p-markings in $\mathcal{Z}$ with $m_2 \in R_{\mathcal{Z}}(m_1)$. The minimum distance of time $d_{\min}(m_1, m_2)$ between m_1 and m_2 is defined as follows:*

$$d_{\min}(m_1, m_2) := \min\{ d_{\min}(z_1, z_2) \mid z_1 \in M_1 \wedge z_2 \in M_2 \}.$$

Definition 3.77 (maximum distance of time between p-markings) *Let $\mathcal{Z}$ be a Time Petri net and m_1 and m_2 reachable p-markings in $\mathcal{Z}$ with $m_2 \in R_{\mathcal{Z}}(m_1)$. The maximum distance of time $d_{\max}(m_1, m_2)$ between m_1 and m_2 is defined as follows:*

$$d_{\max}(m_1, m_2) := \max\{ d_{\max}(z_1, z_2) \mid z_1 \in M_1 \wedge z_2 \in M_2 \}.$$

Remark 3.78 *In every Time Petri net $\mathcal{Z}$ and for p-markings m_1, m_2 and states z_1, z_2 in $\mathcal{Z}$ it holds that:*

1. *For every firing sequence the length of any minimum run and of any maximum run, if it exists, is a natural number.*

2. $d_{\min}(z_1, z_2), d_{\min}(m_1, m_2) \in \mathbb{N}$ *and*
   $d_{\max}(m_1, m_2), d_{\max}(z_1, z_2) \in (\mathbb{N} \cup \{\infty\})$.

Therefore it holds that for every firing sequence there is a minimum run in which all elapsing times are natural numbers. If a maximum run exists then similarly there is a maximum run in which all elapsing times are natural numbers.

Proof:

for 1:

Let $\sigma = t_1 \ldots t_n$ be a firing sequence in $\mathcal{Z}$ and $\sigma(\tau^*)$ with $\tau^* = \tau_0^* \tau_1^* \ldots \tau_n^*$ a minimum run of σ. Let us consider the length $l(\sigma(\tau^*))$ of $\sigma(\tau^*)$.

Assuming that

$$\ell(\sigma(\tau^*)) \notin \mathbb{N}, \tag{76}$$

there is at least one τ_j^*, $0 \leq j \leq n$ with $\tau_j^* \notin \mathbb{N}$. Then, according to Theorem 3.26 there exists a further feasible run $\sigma(\tau^{**})$ of σ such that for all $0 \leq j \leq n$ it holds that $\tau_j^{**} \in \mathbb{N}$, i.e.,

$$\ell\big(\sigma(\tau^{**})\big) \in \mathbb{N}. \tag{77}$$

Hence, because of Theorem 3.26(3) it also holds that

$$\ell\big(\sigma(\tau^{**})\big) =$$
$$\sum_{i=0}^{n} \tau_i^{**} \leq \sum_{i=0}^{n} \tau_i^{*}$$
$$= \ell\big(\sigma(\tau^{*})\big).$$

Thus, (76) and (77) yield

$$\ell\big(\sigma(\tau^{**})\big) < \ell\big(\sigma(\tau^{*})\big),$$

which contradicts the minimality of $\sigma(\tau^{*})$. Hence, $\ell\big(\sigma(\tau^{*})\big) \in \mathbb{N}$.

The corresponding assertion for the length $\ell\big(\sigma(\tau^{*})\big)$ of a maximum run $\sigma(\tau^{*})$ of σ can be proved similarly using Theorem 3.32.

For 2:

Both claims follow similarly to 1. from Theorem 3.26 and Theorem 3.32, respectively.   □

If we discover an undesirable behavior during the analysis of a Time Petri net which models a system, then we would also like to find out *whether* and *how* it is possible to improve the system by changing the time restrictions, first in the model and then in the system. Results allowing us to leave the structure of the system unchanged and avoid the undesirable behavior by changing time restrictions are of course of great practical importance.

3.8.1   Unbounded Time Petri Nets

We now restrict our attention to unbounded Time Petri nets, i.e., nets whose state spaces are infinite. We can hardly use reachability graphs for solving modeling problems in these nets.

In the following we consider several problems for which we present algorithmic solutions. These problems originate from different applications and are used to analyze and to verify specific systems. We will solve the following five problems using systems of linear inequalities or linear programs. All coefficients in these systems of inequalities will be natural numbers but the solutions may be rational or real. The solvability of linear systems of inequalities and of linear programs in $\mathbb{R}$ can be decided in polynomial time, e.g., with the Khachiyan-algorithm (cf. [GLS93] or [PS98]). The solvability in $\mathbb{Q}$ of the problems considered here is also decidable in polynomial time (cf. result of Tardos (1986) for rational polyhedrons in [GLS93]).

For the following five problems we assume all considered transition sequences to be firing sequences in the skeletons of the respective Time Petri nets.

Problem 1

Input:
- A Time Petri net $\mathcal{Z}$,
- a transition sequence σ,
- an arbitrary run $\sigma(\tau^*)$ of σ.

Output:
(1) Is σ is a firing sequence in $\mathcal{Z}$?

(2) If so, compute a feasible run $\sigma(\tau)$ of σ.

(3) Is $\sigma(\tau^*)$ a feasible run in $\mathcal{Z}$?

Solution:

We formulate the parametric run $\sigma(x)$ of σ and the parametric state (z_σ, B_σ).

For (1) It is easy to see that σ is a firing sequence in $\mathcal{Z}$ if and only if the system of linear inequalities B_σ is solvable. To answer the question, it is sufficient to find a solution with real values.

For (2) Every real-valued solution of the system B_σ determines a feasible run of σ.

For (3) $\sigma(\tau^*)$ is a feasible run of σ if and only if τ^* is a solution of B_σ. □

In practical applications of Time Petri nets, as for instance in the area of modeling and analysis of metabolic networks, one needs to allow arbitrary values for the t-marking h_0 of the initial state z_0 so that the initial value of the clock of a transition enabled in z_0 can be a rational number. We will generalize Definition 3.19 to allow for such arbitrary natural numbers or $\sharp$ as the initial values of h_0. In cases where rational initial times would be necessary we scale down the time unit in order to obtain natural numbers for the clocks of all enabled transitions in the initial state. Thus, without loss of generality, the following definitions generalize Definition 3.19:

Definition 3.79

$$h_0(t) := \begin{cases} \sharp & \text{if } t^- \not\leq m_0 \\ c_0^t & \text{otherwise} \end{cases}$$

where all c_0^t are natural numbers.

It is easy to see that the parametric state z_ε should now be defined as follows:

Definition 3.80 $z_\varepsilon = (m_0, h_\varepsilon)$ *with*

$$h_\varepsilon(t) := \begin{cases} \sharp & \text{if } t^- \not\leq m_0 \\ c_0^t + x_0 & \text{otherwise} \end{cases} .$$

Problem 2

Input: • A Time Petri net $\mathcal{Z}$ with an *only* partiallydefined interval function I,

                    • a transition sequence σ.

Output: (1) Can I be extended to a total function such that σ is a firing sequence in the resulting net?

                    (2) If so, compute a feasible run $\sigma(\tau)$ of σ.

Solution:

Let us consider $eft(t)$ and $lft(t)$ of every transition t in $\mathcal{Z}$: If $eft(t)$ is not defined then we associate a variable a_t with t, and similarly if $lft(t)$ is not defined we associate a variable b_t with t. Now we can formulate (z_σ, B_σ) and consider the system of linear inequalities B_σ: The variables in this system are the x_i from the parametric run $\sigma(x)$ as well as all added a_t and b_t. We are looking for a rational solution of the system. As already mentioned, this can be done in time polynomial in $\ell(\sigma)$.

For (1) The system of inequalities B_σ has a solution if and only if the interval function I can be extended to a total function (as described above) such that σ is a firing sequence.

For (2) Each solution computed in (1) obviously gives us a feasible run $\sigma(\tau)$ of σ in $\mathcal{Z}$. $\square$

Problem 3

  Input: • A Time Petri net $\mathcal{Z}$,

                  • a firing sequence σ.

  Output: (1) A minimum run of σ.

                  (2) A maximum run of σ, if it exists.

  Solution:

Let us again consider the system of linear inequalities B_σ and the parametric run $\sigma(x)$:

For (1) Each solution of the linear program

$$\min \left\{ \sum_{i=0}^{\ell(\sigma)} x_i \mid B_\sigma \right\}$$

gives us a minimum run of σ.

For (2) Similarly, Each solution of the linear program

$$\max \left\{ \sum_{i=0}^{\ell(\sigma)} x_i \mid B_\sigma \right\}$$

gives us a maximum run of σ, if there is one. If there is no solution, we know that the set of restrictions B_σ of the linear program has no upper bound and therefore there is no maximum run for σ.

Both linear programs are solvable in time polynomial in $\ell(\sigma)$(cf. [PS98]).

$\square$

Problem 4

Input: • A Time Petri net $\mathcal{Z}$ with an *only* partially defined interval function I,

  • a transition sequence σ,

  • a number $\lambda \in \mathbb{R}_0^+$.

Output: (1) Is it possible to extend I to a total function such that σ is a firing sequence in $\mathcal{Z}$ with a feasible run $\sigma(\tau)$ such that $\ell(\sigma(\tau)) \le \lambda$?

  (2) Is it possible to extend I to a total function such that σ is a firing sequence in $\mathcal{Z}$ with a feasible run $\sigma(\tau)$ such that $\ell(\sigma(\tau)) \ge \lambda$?

Solution:

We first extend I in the same way as for Problem 2 and formulate B_σ.

For (1) We solve in $\mathbb{Q}$ the system of linear inequalities B'_σ consisting of all inequalities of B_σ and the linear inequality $\ell(\sigma(x)) \le \lambda$. The variables of the system are $x_0, \ldots, x_{\ell(\sigma)}$, and all added a_t and b_t.

For (2) The approach is similar to that in (1), but we now consider the linear inequality $\ell(\sigma(\tau)) \ge \lambda$ together with B_σ. $\square$

Problem 5

Input: • A Time Petri net $\mathcal{Z}$ with an *only* partially defined interval function I,

• a transition sequence $\sigma_1 = \sigma t_1$, where σ is a transition sequence and t_1 a transition in $\mathcal{Z}$,

• a transition sequence $\sigma_2 = \sigma t_2$, where t_2 is a transition in $\mathcal{Z}$ with $t_1 \neq t_2$.

Output: Is it possible to extend I to a total function such that σ_1 is a firing sequence in $\mathcal{Z}$ but σ_2 is *not*?

Solution:

First we again extend I as in the solution of Problem 2 and formulate B_σ. Afterwards we formulate the parametric states $z_{\sigma_1} = (\sigma_1(x), B_1)$ and $z_{\sigma_2} = (\sigma_2(y), B_2)$. In order to solve Problem 5 we now have to decide whether the following proper inequality is solvable

$$
\max \Big\{ \underbrace{\sum_{i=0}^{\ell(\sigma_1)} x_i \mid B_1}_{(P_1)} \Big\} < \min \Big\{ \underbrace{\sum_{i=0}^{\ell(\sigma_2)} y_i \mid B_2}_{(P_2)} \Big\}. \tag{78}
$$

We are looking for a solution such that the values of all variables a_{t_i}, b_{t_j} are rational numbers but $x_0, \ldots, x_{\ell(\sigma_1)}$ and $y_0, \ldots, y_{\ell(\sigma_2)}$ can take arbitrary real values.

In order to solve (78) we first consider the linear programs (P_1) and (P_2). The linear program (P_1) can be rewritten in the following form:

$$
\alpha = \max\{a^\mathsf{T} \cdot x \mid A \cdot x \leq b\}
$$

and (P_2) in the form

$$
\beta = \min\{c^\mathsf{T} \cdot y \mid B \cdot y \geq d\}.
$$

We now consider the systems of linear inequalities (U_1) and (U_2), defined as follows:

$$(U_1) \begin{cases} A \cdot x \leq b \\ A^{\mathrm{T}} \cdot u \geq a \\ a^{\mathrm{T}} \cdot x - b^{\mathrm{T}} \cdot u \geq 0 \end{cases} \qquad (U_2) \begin{cases} B \cdot y \geq d \\ B^{\mathrm{T}} \cdot v \leq c \\ d^{\mathrm{T}} \cdot v - c^{\mathrm{T}} \cdot y \geq 0 \end{cases}$$

Using the dual theory for linear programs it is easy to see that a point x^* is optimal for (P_1) if and only if (x^*, u^*) is a solution of U_1 and a point y^* is optimal for (P_2) if and only if (y^*, v^*) is a solution of U_2. Now, we define the linear program (P) as follows:

$$(P) \quad \gamma = \max\{ c^{\mathrm{T}} \cdot y - a^{\mathrm{T}} \cdot x \mid U_1, U_2 \}.$$

The variables in (P) are x, y, u, v as well as all variables a_t and b_t added in order to make the interval function total.

The inequality (78) is solvable if and only if the linear program (P) is solvable and $\gamma > 0$. In this case, an optimal point of (P) determines a possible total definition for I, so we have solved Problem 5 in time polynomial in $\ell(\sigma_1) + \ell(\sigma_2)$. □

Problem 5 answers the question of whether certain time intervals can prevent branching of a firing sequence in a Time Petri net even though the branching is possible in the skeleton. This question can be answered immediately if the transitions t_1 and t_2 have at least one common pre-place. In this case we would define $eft(t_2)$ and $lft(t_1)$ such that $eft(t_2) > lft(t_1)$ if possible (if at least one of them is not yet fixed in the beginning). Otherwise, as mentioned above, the solution is more complicated.

3.8.2   Bounded Time Petri Nets

In this section we consider bounded Time Petri nets. For such nets we can solve problems which require knowledge about the whole state space of the net.

Before we consider the next four problems we introduce some new notions:

Let $\mathcal{Z} = (P, T, F, V, m_0, I)$ be an arbitrary Time Petri net. Let $\mathcal{RG}_{\mathcal{Z}} = (W, E, T \cup (\mathbb{N} \times T))$ be the reachability graph of $\mathcal{Z}$. This graph is an edge-labeled digraph with labels from the set $T \cup (\mathbb{N} \times T)$, i.e., each edge $k \in E$ is assigned an edge-label $\kappa_k = n, t$ or $\kappa_k = t$ (for $n = 0$). Using this labeling we now define a weight-function $w : E \longrightarrow \mathbb{N}$ as follows:

$$w(k) := \begin{cases} 0 & \text{if } \kappa_k = t \\ n & \text{if } \kappa_k = n, t \end{cases} \quad \text{for each } k \in E.$$

Thus, for each Time Petri net $\mathcal{Z}$ we obtain the weighted digraph $\mathcal{RG}_{\mathcal{Z}}^w :=$ (W, E, w) from its reachability graph $\mathcal{RG}_{\mathcal{Z}} = (W, E, T \cup (\mathbb{N} \times T))$.

We will use the graph $\mathcal{RG}_{\mathcal{Z}}^w := (W, E, w)$ to compute the minimum and maximum distance of time from one state to another or from a p-marking to a second one.

The graph algorithms we use all run in time linear or polynomial in the size of $\mathcal{RG}_{\mathcal{Z}}^w$ and thus, also in the size of the reachability graph $\mathcal{RG}_{\mathcal{Z}}$.

The basic notions from graph theory as well as basic graph algorithms which are not introduced here can be found in "Introduction to Algorithms" by Cormen, Leiserson, Rivest, and Stein (cf. [CLRS01]).

Problem 6

Input: • A Time Petri net $\mathcal{Z}$,

   • integer-states z_1 and z_2, reachable in $\mathcal{Z}$.

Output: (1) Is there is a firing sequence from z_1 to z_2?

   (2) If $z_2 \in RS_{\mathcal{Z}}(z_1)$, compute the minimum distance of time from z_1 to z_2 as well as a minimum run from z_1 to z_2 realizing this minimum distance.

Solution:

Let us consider the states z_1 and z_2. Because both states are reachable and integer-states they are vertices in the reachability graph $\mathcal{RG}_{\mathcal{Z}}$ and therefore also in the graph $\mathcal{RG}_{\mathcal{Z}}^w$.

Part (1) of the problem is solved by answering whether there exists a path from z_1 to z_2 in $\mathcal{RG}_{\mathcal{Z}}^w$.

Because of Remark 3.78, part (2) is solved by computing a shortest path in the weighted digraph $\mathcal{RG}_{\mathcal{Z}}^w$.

Therefore because all weights are natural numbers both parts of the problem can be solved using, e.g., the Dijkstra algorithm. The algorithm either finds a path from z_1 to z_2 or determines that there is none in polynomial time (cf. *shortest-path algorithm* in [CLRS01]). The minimum distance of time from z_1 to z_2 is now the sum of all weights along the found shortest path.   □

Problem 7

Input: • A Time Petri net $\mathcal{Z}$,

            • p-markings m_1 and m_2, reachable in $\mathcal{Z}$.

Output: (1) Is $m_2 \in R_{\mathcal{Z}}(m_1)$?

            (2) If $m_2 \in R_{\mathcal{Z}}(m_1)$, compute the minimum distance of time from m_1 to m_2 as well as a minimum run realizing this.

Solution:

Similarly as in the solution of the previous problem we consider the graph $\mathcal{RG}_{\mathcal{Z}}^w$. Furthermore we define the following sets of vertices M_1 and M_2:

$$M_i := \{\, z \mid z = (m_i, h) \land z \in RIS_{\mathcal{Z}} \,\} \quad \text{for } i = 1, 2.$$

The questions (1) and (2) now obviously ask about the existence of a shortest path from some element of M_1 to an element of M_2 in the graph $\mathcal{RG}_{\mathcal{Z}}^w$. This problem can also be solved in time polynomial in the size of the graph (cf. *all-pairs shortest paths algorithm* in [CLRS01]). Thereby we obtain the values d_{ij} for $i, j = 1, \ldots, |RIS_{\mathcal{Z}}|$ with

$$d_{ij} = \begin{cases} d_{\min}(z_i, z_j) & \text{if } z_2 \in RS_{\mathcal{Z}}(z_1) \\ \infty & \text{otherwise} \end{cases}.$$

Clearly, the value $d_{\min}(m_1, m_2)$ is now the minimum of a finite set of values d_{ij}:

$$d_{\min}(m_1, m_2) = \min \{ d_{ij} \mid z_i \in M_1 \land z_j \in M_2 \}. \qquad \square$$

Problem 8

Input:
- A Time Petri net $\mathcal{Z}$,
- integer-states z_1 and z_2, reachable in $\mathcal{Z}$.

Output:
(1) Is there a firing sequence from z_1 to z_2?

(2) If $z_2 \in RS_{\mathcal{Z}}(z_1)$, compute the maximum distance of time from z_1 to z_2 as well as a maximum run realizing this.

Solution:

We again consider $\mathcal{RG}^w_{\mathcal{Z}} = (W, E, w)$ (cf. p. 129). Using this graph we define the graph $\mathcal{G}_{\mathcal{Z}} := (W', E', w')$ with

$$W' := W,$$
$$E' := E \setminus \{ (z_2, z) \mid (z_2, z) \in E \},$$
$$w' := -w_{|def(E')}.$$

The graph $\mathcal{G}_{\mathcal{Z}}$ is obtained from $\mathcal{RG}^w_{\mathcal{Z}}$ by deleting all output edges of z_2 and negating all weights $w(k)$, i.e., $w'(k) := -w(k)$. Thus, $\mathcal{G}_{\mathcal{Z}}$ is also a weighted digraph. Deleting all output edges of z_2 ensures that any path containing this vertex ends once the vertex is reached (cf. Definition 3.75).

The weights are negated in order to use a shortest-path algorithm in $\mathcal{G}_{\mathcal{Z}}$ to find a longest[10] path from z_1 to z_2 in $\mathcal{RG}^w_{\mathcal{Z}}$. For finding such a path however, we can no longer use the Dijkstra algorithm because the weights are now non-positive integers. Thus, we use the Bellman-Ford algorithm which is slower than the Dijkstra algorithm but also solves the task in polynomial time.

We present here another algorithm which computes the largest path in $\widetilde{\mathcal{G}}_{\mathcal{Z}} := (W', E', w_{|def(E')})$ (in which all output edges of z_2 are deleted) in linear time and therefore solves (2) as well as (1):

---

[10]For an arbitrary linear program $\max\{ f(x) \mid x \in L \}$ it always holds that: $\max \{ f(x) \mid x \in L \} = - \min \{ -f(x) \mid x \in L \}$.

(1) Compute the strongly connected components of $\widetilde{\mathcal{G}}_{\mathcal{Z}}$.

(2) For the component $Q_{z_1} := (V_{z_1}, E_{z_1})$ containing z_1 check whether $|V_{z_1}| \geq 2$. If **yes** then go to (6), if **no** then go to (3).

(3) Compute the acyclic graph $\widetilde{\mathcal{G}}_{\mathcal{Z}}^{SCC}$ of the strongly connected components of $\widetilde{\mathcal{G}}_{\mathcal{Z}}$.

(4) Check in $\widetilde{\mathcal{G}}_{\mathcal{Z}}^{SCC}$ whether it is possible starting in the vertex Q_{z_1} to reach a vertex $P := (V_P, E_P)$ such that $|V_P| \geq 2$.
     If **yes** then go to (6), if **no** then go to (5).

(5) Compute $d_{\max}(Q_{z_1}, Q_{z_2})$. $d_{\max}(z_1, z_2) := d_{\max}(Q_{z_1}, Q_{z_2})$. STOP.

(6) $d_{\max}(z_1, z_2) := \infty$. STOP.

This algorithm obviously runs in linear time.   □

Problem 9

Input: • A Time Petri net $\mathcal{Z}$,

                • p-markings m_1 and m_2, which are reachable in $\mathcal{Z}$.

Output: (1) Is $m_2 \in R_{\mathcal{Z}}(m_1)$?

                (2) If $m_2 \in R_{\mathcal{Z}}(m_1)$, compute the maximum distance of time from m_1 to m_2 as well as a maximum run realizing this.

Solution:

Similarly as for Problem 7 we consider the sets of vertices M_i, $i = 1, 2$. An algorithm computing the *all-pairs shortest path* in the weighted digraph $\mathcal{G}_{\mathcal{Z}}$ obviously gives us a solution to the problem.   □

Finally, we note that the minimum and maximum distance of time from an integer-state to a p-marking and from a p-marking to an integer-state can also be computed. It is obvious that these four problems can be solved in polynomial or linear time. Computing the minimum and maximum distance of time from an integer-state to a p-marking is equivalent to the *single-source shortest path problem*. The *single-destination shortest path problem* solves the task of finding the minimum and maximum distance of time from a p-marking to an integer-state. Both algorithms run in polynomial time, cf. [CLRS01].

3.9 Bibliographical Notes

Time Petri nets were first introduced by Merlin in [Mer74] and the first algorithm for the analysis of these nets was presented by Berthomieu and Menasche in [BM83] and extensively studied by Berthomieu and Diaz in [BD91]. A slight variation of this algorithm can be found in some articles by Berthelot and Boucheneb, as, for instance, in [BB94] and [BB93].

The method presented by Berthomieu and Menasche is based on the definition of reachability graphs whose vertices are state classes. The reachability graph as well as the notion of a state class are defined differently than in this book. Every edge of the graph represents the firing of a transition. The elapses of times are not explicitly shown in the graph. A class as defined by Berthomieu and Menasche is a pair consisting of a marking (in the sense of place-markings) and a domain reflecting the time situation in the net. The domain associates a variable with each enabled transition in the considered marking. The values of these variables represent possible points in time for the firing of this transition. Thus, a set of possible points in time (interval) for firing is defined for each enabled transition. All the sets in the considered marking together define the domain of the class. A domain is defined by linear inequalities, i.e. a domain is a system of linear inequalities. An advantage of this definition of state classes over the definition presented here is that the number of variables in a certain class is exactly the same as the number of enabled transitions in the marking of the class. A disadvantage is the possibly large number of inequalities defining the class. Hence, in order to compute a vertex in the reachability graph introduced by Berthomieu and Menasche, one always has to compute the bounds for the variables in order to formulate the system of inequalities.

The algorithm reducing the state space of an arbitrary Time Petri net to the set of all reachable integer-states or all reachable essential-states presented in this book is based on the the notion of *parametric states*. An earlier version of this reduction was done for finite Time Petri nets in [Pop91] (in German in [PZ89]) using an imaginary global clock. It should be noted that the set of time conditions B_σ in the notion of a *parametric state* (z_σ, B_σ) can be formulated as a set of formulas of first-order logic (or even of Presburger arithmetic) (cf. [PZS99] and [PZ07]).

A series of examples comparing the size of the two kinds of reachability graphs ("state-class"-reachability graphs and "essential-state"-reachability graphs) is given in [Pil09] (see also Appendix). For nets with high concurrency the "essential-state"-reachability graph tends to be smaller than the respective "state-class"-reachability graph.

A disadvantage of the "state-class"-reachability graph compared to the "essential-state"-reachability graph is the fact that time is not explicitly represented and hence, a quantitative analysis using these graphs is hardly possible. However, [BFSV04] presents an algorithm which computes the minimum and maximum lengths of time for a given firing sequence in the Time Petri net. The solution to Problem 3 is a stronger result than this, since Problem 3 allows firing as well as non-firing sequences as input.

Furthermore, a transformation of a Time Petri net into a timed automaton (cf. [AD94]) was presented by Cassez and Roux in [CR04]. The transformation is rather costly because a timed automaton is associated with each transition. For some quite strongly restricted classes of Time Petri nets an easier transformation was found in [Pen00] and [PP06]. The goal of such transformations is use algorithms developed for timed automata in the analysis of Time Petri nets. The obvious downsides to this approach are the size of the resulting timed automaton and the difficult transfer of information back into the net after the analysis of the automaton. Therefore, algorithms and tools developed specifically for Time Petri nets are of great practical importance.

3.10 Exercises

Exercise 3.1

Let the number-theoretical function $f(x_1, x_2)$ be defined as follows

$$f(x_1, x_2) := \begin{cases} 2 \cdot x_1 - x_2 & \text{if } 2 \cdot x_1 \geq x_2 \\ \text{undefined} & \text{otherwise} \end{cases}.$$

Give a counter machine and the corresponding Time Petri net computing f.

Exercise 3.2

Let us consider the Time Petri net $\mathcal{Z}_3$:

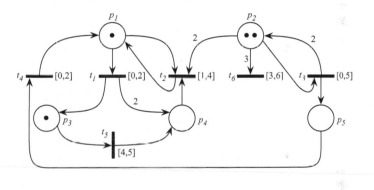

Give the parametric state (z_σ, B_σ) for $\sigma = t_3 t_6 t_1$.

Exercise 3.3

Let $\mathcal{Z} = (P, T, F, V, m_0, I)$ be a Petri net. Furthermore let (z_σ, B_σ) be a parametric state and $\sigma = t_1 \cdots t_n$ a transition sequence in $\mathcal{Z}$. The number of variables appearing in B_σ is at most $n + 1$. Show that the number of non-redundant inequalities in B_σ is at most $\min\{2 \cdot (n \cdot |T| + 1), (n + 1) \cdot (\frac{n}{2} + 2)\}$.

Exercise 3.4

Give a feasible run $\sigma(\hat{\beta})$ in a Time Petri net $\mathcal{Z}$ such that

$$1 \leq \left[\!\left[\sum_{k=0}^{n} x_k \right]\!\right]_{\hat{\beta}} - \left[\!\left[\sum_{k=0}^{n} x_k \right]\!\right]_{\beta^*},$$

where β^* and $\hat{\beta}$ are defined as in the proof of Theorem 3.26.

Exercise 3.5

Consider Theorem 3.32: Give a recursive definition of the assignment β^*.

Exercise 3.6

Give a proof for the following generalization of Theorem 3.61: Let $\mathcal{Z} = (P, T, F, V, m_0, I)$ be a Time Petri net with

(a) $S(\mathcal{Z})$ is an AC-net,[11]

(b) $S(\mathcal{Z})$ is homogeneous,

(c) for each place $p \in P$ it holds that: $\mathcal{M}in(p) \leq \mathcal{M}ax(p)$, and

(d) for each transition $t \in T$ it holds that: $lft(t) > 0$.

Then if $S(\mathcal{Z})$ is live $\mathcal{Z}$ is live, too.

Hint: A solution to this exercise can be found in [Bac11] and in [BPZ10].

Exercise 3.7

Show that the Petri net $\mathcal{N} := (P, T, F', V', m_0)$, obtained from $S(\mathcal{Z}) = (P, T, F, V, m_0)$ after Step 1 in the proof of Theorem 3.65 is an EFC net.

Exercise 3.8

---

[11]In this book AC nets are defined for arbitrary Petri nets, cf. Definition 2.19.

Give a Time Petri net $\mathcal{Z}$ satisfying the conditions $(a), (b), (c)$ and (d) of Theorem 3.65 but not condition (e), such that the net $\mathcal{Z}'$ (defined in Step 1 in the proof of Theorem 3.65) does not comply with all conditions of Theorem 3.61.

Remark: With this exercise the necessity of assumption (e) for the proof of Theorem 3.65 is shown.

Exercise 3.9

The firing of a transition in a Time Petri net is defined using the notion of "static conflict", cf. Definition 3.7, i.e., if two transitions with a common pre-place are enabled and one of them fires, then if the second transition is still enabled, its clock is reset to zero in the state after the firing.

(a) Give a definition using the notion of "dynamic conflict", i.e., if two transitions with a common pre-place are enabled and one of the transitions fires, then if the second transition is still enabled, its clock is not reset but continues measuring time.

(b) Do Theorems 3.26 and 3.32 hold if transitions fire according to this new definition?

Chapter 4

Timed Petri Nets

In this chapter we consider a further important class of time-dependent Petri nets, the Timed Petri nets. In any real system every event takes some amount of time, no matter how small. In Timed Petri nets this fact is incorporated by associating a duration with each transition. To model a real system with a Timed Petri net the duration of a transition is set to the duration of the event or process modeled by this transition. Therefore, in most studies on Timed Petri nets, durations of length zero are not considered, but allowing time durations of certain events to be zero can be very useful for practical reasons. If, for instance, we want to model events of which one has a very small time duration compared with the others, then it might be convenient to set the small duration to zero, in order to avoid using huge values for the durations of the other events. Later on, this simplifies the analysis of the net.

Through the way in which time is attributed to transitions and because the maximal-step rule is used for firing it is exactly determined when events take place and the concurrency of the timeless skeleton of such a net is lost. Timed Petri nets are however a very useful means of modeling because of their simplicity and clearness but also because these nets and their analysis and verification have already been thoroughly studied. They are also well suited for combination with other types of Petri nets so that various extensions of Timed Petri nets such as Interval-Timed Petri nets (cf. [PZP12]) and Timed Petri nets with priorities (cf. [WPZR04], [RPZW02]) have been developed as intuitive modeling tools for a wide variety of real systems.

In the following we introduce Timed Petri nets, in which zero durations are allowed, and also reachability graphs for this class of nets, and show their computational equivalence to Turing machines. After that we explore the possibility of a local transformation of Timed Petri nets into Time Petri nets. We finally define time-dependent state equations for Timed Petri nets consistently with the respective notion for classic Petri nets and show a sufficient condition for non-reachability of place-markings.

4.1   Definitions

Timed Petri nets are classical Petri nets in which a time duration d_t is associated with each transition t. The firing of the transition t now takes exactly d_t time units. Transitions are fired according to the *maximal-step* firing rule, i.e., in each marking a maximal set of enabled transitions fires at once, immediately after their enabling. The time durations are generally rational numbers.

Definition 4.1 (Timed Petri net) *A Timed Petri net is a 6-tuple* $\mathcal{D} = (P, T, F, V, m_0, D)$ *such that*

    1. $S(\mathcal{D}) = (P, T, F, V, m_0)$ is a Petri net and

    2. $D : T \longrightarrow \mathbb{Q}_0^+$.

The function D is called the *duration function* of the net $\mathcal{D}$. Fig. 4.1 shows a Timed Petri net. In the graphical representation of a Timed Petri net the time duration of each transition is written next to the transition in angle brackets.

We call the classic (timeless) Petri net $S(\mathcal{D})$ the *skeleton* of $\mathcal{D}$. Similarly as for Time Petri nets, it is easy to see that without loss of generality we can represent any rational time durations in a specific net by natural numbers. Therefore, as of now, we assume the domain of every duration function to be $\mathbb{N}$.

Again similarly as for Time Petri nets, the (place-)markings alone cannot describe the behavior of a Timed Petri net. A state needs to contain a place-marking as well as a transition-marking and these two markings together fully describe the situation in a Timed Petri net.

Example 4.2

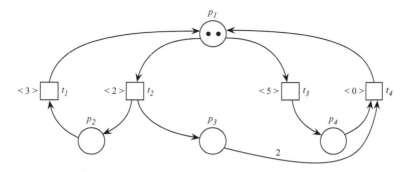

Figure 4.1: The Timed Petri net $\mathcal{D}_1$

Definition 4.3 (state) *A state in the Timed Petri net* $\mathcal{D} = (P, T, F, V, m_0, D)$
is a pair $z = (m, u)$ *such that*

1. *m is a marking in $S(\mathcal{D})$ and*

2. *$u : T \longrightarrow \mathbb{R}_0^+$ with $\forall t \, (\, (t \in T \wedge t^- \leq m) \longrightarrow u(t) \leq D(t) \,).$*

We call m a *place-marking* (short: p-marking) and u a *transition-marking*
(short: t-marking) of $\mathcal{D}$. We understand $u(t)$ to be the clock of transition t.
The transition clocks in Timed Petri nets work differently from the transition
clocks in Time Petri nets: They count down the remaining time until their
transition has finished firing. Therefore, as the *initial state* of a Timed Petri
net $\mathcal{D} = (P, T, F, V, m_0, D)$ in which no transition is in the process of firing
we define the state $z_0 := (m_0, u_0)$ with $u_0(t) := 0$ for all $t \in T$.

In order to define the firing mode we formally introduce the notion of a
maximal step:

Definition 4.4 (maximal step) *Let* $z = (m, u)$ *be a state in the Timed
Petri net* $\mathcal{D} = (P, T, F, V, m_0, D)$. *Then* $M \subseteq T$ *is a maximal step in* z, *if*

(1) $\forall t \, (\, t \in M \longrightarrow u(t) = 0 \,),$

(2) $\sum\limits_{t \in M} t^- \leq m$,

(3) $\forall \hat{t} ((\hat{t} \in T \wedge \hat{t} \notin M \wedge \hat{t}^- \leq m \wedge u(\hat{t}) = 0) \longrightarrow (\sum\limits_{t \in M} t^- + \hat{t}^-) \not\leq m)$.

The first two conditions in Definition 4.4 define the notion of a *step*. A step is a set of transitions. The first condition ensures that none of these transitions is in the process of firing and the second condition states that not only is every transition in M enabled in m but the pre-places of the transitions in M hold enough tokens for all transitions in M to fire at once. In some books this property is called *collective enabledness*. The third condition disallows the existence of a proper superset of M which fulfills the previous two conditions.

It is easy to see that the notion of a *maximal step* can also be defined time-independently by omitting condition (1) and omitting $u(\hat{t}) = 0$ in condition (3). Thus, we can use maximal-step firing mode in timeless Petri nets, too, even if this is unusual. Timeless Petri nets firing in maximal-step mode have more expressive power than our classic timeless Petri nets and are even Turing equivalent. This can be proved by showing the equivalence between this kind of nets and counter machines. We have introduced counter machines in Section 3.3. The first three commands *start, stop* and *INC* can obviously be simulated by Petri nets firing in maximal steps because in our simulation of these commands by classic Petri nets there was always exactly one transition enabled (cf. p.41 and p.41, respectively). We still need to prove that the fourth command $l : DEC(i) : r : s$ (cf. p.39), the so-called *zero-test*, can also be simulated by such a timeless net.

The command DEC is simulated by the Petri net presented in Fig. 4.2. Assuming that the place w_i is empty, only the sequence of maximal steps $\sigma_1 = M_1^1 M_2^1 M_3^1$ with

$$M_1^1 = \{t_3\}, M_2^1 = \{t_4\}, M_3^1 = \{t_5\}$$

can fire in $\mathcal{N}_2$. After that, the place p_r contains a token and the place p_s is empty. Otherwise, assuming that the place w_i is marked, only the sequence of maximal steps $\sigma_2 = M_1^2 M_2^2 M_3^2$ with

$$M_1^2 = \{t_3\}, M_2^2 = \{t_1, t_4\}, M_3^2 = \{t_2\}$$

can fire in $\mathcal{N}_2$. After that, the place p_s contains a token and the place p_r is empty. Thus, $\mathcal{N}_2$ simulates the command $l : DEC(i) : r : s$ and the Turing equivalence of timeless Petri nets firing in maximal-step mode follows.

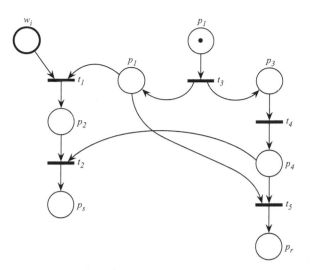

Figure 4.2: Simulation of the command $l : DEC(i) : r : s$ by a timeless Petri net firing in maximal-step mode

We will now define the rules for state change in Timed Petri nets: one rule determining when an enabled transition starts to fire and one for the elapsing of time. As already mentioned, similarly to Time Petri nets, we can interpret $u(l)$ as a clock associated with the transition t, here however, as a timer, counting down the time until t has finished firing. In every state $z = (m, u)$ the clock $u(t)$ shows whether the transition is in the process of firing (is active) or is not active. Whenever a transition t is not in the process of firing it holds that $u(t) = 0$. This does not mean that t is disabled in m. When a transition starts to fire then its clock $u(t)$ is set to $D(t)$. With time elapsing the value of $u(t)$ is decremented. Hence, the clock $u(t)$ shows the time remaining in z until t reaches the end of its current firing. There is an ambiguity when a transition t has duration zero, i.e., $D(t) = 0$. In this case the value of the clock may indicate that the transition t starts to fire in the state $z = (m, u)$, that it ends its firing, or that it is disabled in m. This is no problem since the enabledness of any transition in a state can be determined using the place-marking.

We define Timed Petri nets in such a way that a firing process once started cannot be interrupted or stopped until it has ended.

1st rule for state change:

Definition 4.5 (firing) *Let* $z_1 = (m_1, u_1)$ *be a state in the Timed Petri net* $\mathcal{D} = (P, T, F, V, m_0, D)$ *and* $M \subseteq T$. *Then* M *can fire in* z_1 *(notation:* $z_1 \xrightarrow{M}$) *if* M *is a maximal step in* z_1. *After the firing of* M *the net* $\mathcal{D}$ *changes into the state* $z_2 = (m_2, u_2)$ *(notation:* $z_1 \xrightarrow{M} z_2$) *with:*

(1) $m_2 := m_1 - \sum\limits_{t \in M} t^- + \sum\limits_{\substack{t \in M, \\ D(t)=0}} t^+,$

(2) $u_2(t) := \begin{cases} D(t) & \text{if } t \in M \\ u_1(t) & \text{otherwise} \end{cases}.$

2nd rule for state change:

Definition 4.6 (elapsing of time) *Let* $z_1 = (m_1, u_1)$ *be a state in the Timed Petri net* $\mathcal{D} = (P, T, F, V, m_0, D)$. *Then, the elapsing of one time unit is possible in* $\mathcal{D}$ *(notation:* $z_1 \xrightarrow{1}$), *if*

$$\forall t \left((t \in T \wedge u_1(t) = 0) \longrightarrow t^- \not\leq m_1 \right).$$

After the elapsing of one time unit the Timed Petri net $\mathcal{D}$ *is in the state* $z_2 = (m_2, u_2)$ *(notation:* $z_1 \xrightarrow{1} z_2$) *with:*

(1) $m_2 := m_1 + \sum\limits_{\substack{t \in T, \\ u_1(t)=1}} t^+,$

(2) $u_2(t) := \begin{cases} u_1(t) - 1 & \text{if } u_1(t) \geq 1 \\ 0 & \text{otherwise} \end{cases}.$

The conditions in Definition 4.6 ensure that firing of transitions is prioritized over the elapsing of time and that a maximal set of enabled transitions in the step immediately start to fire.

The state change by elapsing time can also be defined for arbitrary $\tau \in \mathbb{R}_0^+$ instead of one time unit. Such a definition does not change the behavior of the net, because the "relevant states", i.e., states in which the place-marking changes, are always integer-states. This is obvious because when a transition starts to fire, its clock is set to a natural number and thus the amount of time units after which the firing ends is also a natural number.

The notions of *reachable states, dead states, dead p-markings, boundedness, liveness*, etc. for Timed Petri nets are defined as usual. Likewise similarly to Time Petri nets, the notion of the reduced reachability graph $\mathcal{RG}_\mathcal{D}$ of a Timed Petri net $\mathcal{D}$ is defined. However, we now "fuse together" input edges labeled with maximal steps and output edges labeled with time – this results from the fact that in Timed Petri nets, a step generally needs to fire before time can elapse.

Example 4.7
In Fig. 4.3 part of the reachability graph $RG_{\mathcal{D}_1}$ of the Timed Petri net $\mathcal{D}_1$ is shown. The states z_i for $i = 0, \ldots, 11$ are defined as follows:

$$
\begin{aligned}
z_0 &= \big((2,0,0,0),(0,0,0,0)\big) & z_1 &= \big((0,1,1,0),(0,0,3,0)\big) \\
z_2 &= \big((1,0,1,1),(0,0,0,0)\big) & z_3 &= \big((0,0,1,2),(0,0,0,0)\big) \\
z_4 &= \big((0,1,2,1),(0,0,0,0)\big) & z_5 &= \big((1,0,0,0),(3,0,0,0)\big) \\
z_6 &= \big((1,0,0,0),(0,0,2,0)\big) & z_7 &= \big((0,1,1,1),(0,0,0,0)\big) \\
z_8 &= \big((1,1,1,0),(0,0,0,0)\big) & z_9 &= \big((1,0,1,0),(0,0,2,0)\big) \\
z_{10} &= \big((1,1,2,0),(0,0,0,0)\big)
\end{aligned}
$$

The state z_3 is a dead state. The place p_3 is not bounded and therefore, the reachability graph is infinite.

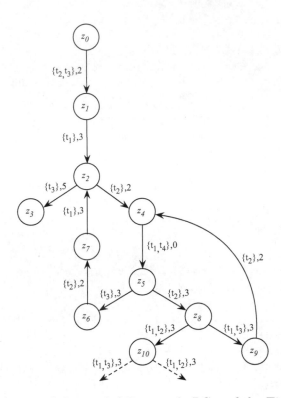

Figure 4.3:   A part of the reachability graph $RG_{\mathcal{D}_1}$ of the Timed Petri net $\mathcal{D}_1$

4.2   Timed Petri Nets and Counter Machines

The considerations in the previous section immediately indicate that counter machines can be simulated by Timed Petri nets and therefore that Timed Petri nets are Turing equivalent. For the sake of completeness, we show how the four commands of counter machines can be simulated by Timed Petri nets (cf. p.39). As with Time Petri nets we model each number l of an assignment by place p_l. Such a place holds a token whenever the considered assignment is being executed. Each counter K_i is modeled by a place w_i. For each step in the computation (program execution) it will hold that $m(w_i) = K_i$ for the p-marking m corresponding to that computation step.

We model the commands by Timed Petri nets as indicated in Table 4.1 and Table 4.2.

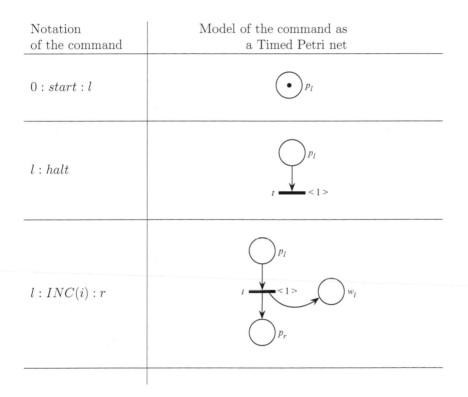

Notation of the command	Model of the command as a Timed Petri net

Table 4.1: Translation of the four possible commands of a counter machine into Timed Peri nets (modules)

Notation of the command	Model of the command as a Timed Petri net
$l : DEC(i) : r : s$	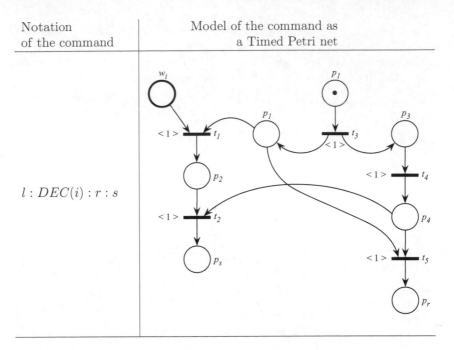

Table 4.2: Continuation of the translation of the four possible commands of a counter machine into Timed Peri nets (modules)

The notion of *Timed PN-computability* can be introduced similarly to *TPN-computability* and *PN-computability*. A number-theoretical function is then Timed PN-computable if and only if it is TPN-computable.

4.3   Transformation of a Timed Petri net into a Time Petri net

As mentioned in the previous section, Timed Petri nets and Time Petri nets have the same expressive power. In the following we will present a *local* transformation, which can transform most Timed Petri nets into Time Petri nets. This transformation is very useful especially for slightly modified Timed Petri nets, where the duration of transitions is not defined exactly but varies (cf. Section 4.5.2). The limitation on the transformation is that the codomain of the duration function D is the set of all positive integers, i.e., the duration zero is not allowed. The limitation results from the different firing modes of the two kinds of time-dependent Petri nets.

Let $\mathcal{D} = (P, T, V, m_0, d)$ be a Timed Petri net. We obtain the Time Petri net $\mathcal{Z_D}$ from the Timed Petri net $\mathcal{D}$ as follows:

We let $\mathcal{Z_D} := (P \cup \{A_t, B_t \mid t \in T\}, \{t^*, t^{**} \mid t \in T\}, F^*, V^*, m_0^*, I)$ and F^*, V^*, m_0^* and I are defined as indicated in Fig. 4.4. We refrain from giving a more formal definition here.

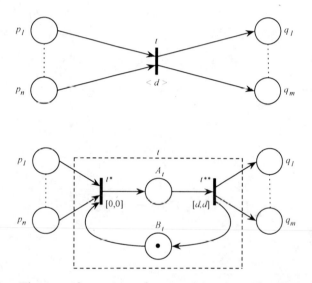

Figure 4.4:   The transformation of a transition in a Timed Petri net to a Time Petri net

The sets of all reachable place-markings in $\mathcal{D}$ and in $\mathcal{Z_D}$ can easily be derived from each other. The same is true for the time relations in both nets.

The following example shows how the transformation is applied to a Timed Petri net including a transition with time duration zero. In this case the sets of reachable p-markings differ.

Example 4.8

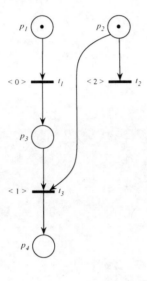

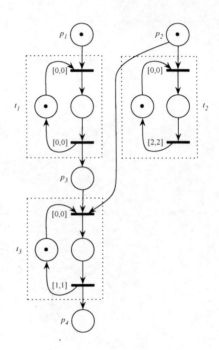

Figure 4.5: The Timed Petri net $\mathcal{D}_2$

Figure 4.6: The Time Petri net $\mathcal{Z}_{\mathcal{D}_2}$

In the Timed Petri net in Fig. 4.5 only the firing of the maximal step $\{t_1, t_2\}$ is possible. The transition t_3 is dead. The p-marking m with $m(P_1) = m(P_4) = 0, m(P_2) = m(P_3) = 1$ is not reachable in $\mathcal{D}$. The transformed *net is the Time Petri net $\mathcal{Z}_{\mathcal{D}_2}$, in which the p-marking m with $m(P_1) = m(P_4) = 0, m(P_2) = m(P_3) = 1$ is reachable and the transition t_3 can fire.*

It is obvious that these nets have different behavior. There are however Timed Petri nets including transitions with time duration zero that have the same behavior as their transformed Time Petri nets.

4.4 State Equations

For the analysis of bounded Timed Petri nets we generally use the reachability graph but when the reachability graph is very large or the net itself is unbounded, we need other methods of analysis.

For classic Petri nets we introduced the notion of the state equation of a firing sequence σ, where $m, C_\mathcal{N}$ and π_σ are given and m' is computed:

$$m' = m + C_\mathcal{N} \cdot \pi_\sigma. \tag{1}$$

We can however also consider the two markings and the incidence matrix as given and try to find a Parikh vector π_σ such that the equation (1) holds. In this case we obtain an equation (a system of equalities) whose variables are the components of π_σ. If equation (1) is not solvable in $\mathbb{N}$ then it is clear that the marking m' is not reachable from m in the considered net, but this is only a sufficient condition. Equation (1) might be solvable in $\mathbb{N}$ but if none of the solutions is the Parikh vector of some firing sequence in the net m' is still not reachable from m. Furthermore, it is clear that the set of all reachable p-markings in a Timed Petri net is a subset of the set of all reachable markings of its skeleton. Therefore, applying the method of state equations in order to check the reachability of a marking from another one in the skeleton is not very useful.

In this section we will extend the notion of the state equation of a firing sequence σ for Timed Petri nets respecting the time. The extension is done consistently. Thus, using *time-dependent state equations* we can exclude the existence of undesirable situations in the system modeled by a Timed Petri net.

For this reason we will classify all tokens in every place – we will distinguish between immediately available tokens, tokens available after one time unit, tokens available after two time units, etc. We realize this by defining *time-extended place markings* (short: *time p-markings*) as well as some auxiliary notions: By $\mathcal{M}(k, l)$ we denote the set of all matrices over $\mathbb{N}$ with k rows and l columns. For a matrix $A \in \mathcal{M}(k, l)$ the notation $A_{.i}$ stands for the i-th column and A_j for the j-th row in A. All the new notions are exemplified using the Timed Petri net $\mathcal{D}_4$, given in Fig. 4.7.

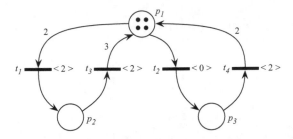

Figure 4.7: The Timed Petri net $\mathcal{D}_3$

Definition 4.9 (time dimension) *Let $\mathcal{D} = (P, T, F, V, m_o, D)$ be a Timed Petri net. The time dimension of $\mathcal{D}$ is the number $d_\mathcal{D} := \max\limits_{t \in T}\{D(t)\} + 1$.*

For the sake of simplicity we write d instead of $d_\mathcal{D}$ whenever $\mathcal{D}$ is clear from the context.

The time dimension of $\mathcal{D}_3$ is $d = 3$.

Subsequently, we will denote the duration $D(t_k)$ of a transition t_k by d_k.

Definition 4.10 (time-extended p-marking) *Let $\mathcal{D} = (P, T, F, V, m_o, D)$ be a Timed Petri net and $d_\mathcal{D} = d$ its time dimension. Then a time-extended place-marking (short : time p-marking) in $\mathcal{D}$ is a matrix $m \in \mathcal{M}(|P|, d)$.*

The idea of a time-extended place-marking is the following: Rather than a vector, this marking is a matrix, in which the first column specifies for each place the number of tokens available in the current state. These are the tokens that can be used for firing in this state. The second column contains the number of tokens that will appear in the place after the elapsing of one time unit and similarly for all the columns of m. These are the tokens that transitions which are currently firing will put on the places once they finish firing. The time p-marking is thus a matrix which has a row for each place of the net and d columns, that is, the matrix has enough columns to keep track of all tokens that will be released by transitions currently in the process of firing and the i-th column contains the tokens that will be released after $i-1$ time units.

The *initial time p-marking* is the matrix $m^{(0)} \in \mathcal{M}(|P|, |T|)$ with $m_{.1}^{(0)} = m_0$ and $m_{ij}^{(0)} = 0$ for $i = 1, \ldots, |P|$ and $j = 2, \ldots, d$.

For the Timed Petri net $\mathcal{D}_3$ the initial time p-marking is

$$
m^{(0)} = \begin{pmatrix} 4 & 0 & 0 \\ 0 & 0 & 0 \\ 0 & 0 & 0 \end{pmatrix}.
$$

So in addition to modeling the current moment a time state also contains information about the future. Note however that the time state cannot fully predict how many tokens will appear in the places, since additional transitions with a shorter duration may fire and release tokens before the transitions with the longest durations release their tokens (the latest releases of tokens predicted by the marking) (cf. Definition 4.11). Furthermore, it can easily be seen that the notion of a time p-marking is a consistent refinement of the notion of a p-marking. The pair $z = (m, u)$ of a time p-marking m and a t-marking u is a *time-extended state* (short: *time state*). We can now redefine the rules for time state change as well as the notion of a maximal step for time states. We do this consistently with Definition 4.4 of maximal steps and the firing rules for Timed Petri nets introduced in Definitions 4.5 and 4.6, replacing the p-marking m by $m_{.1}$.

We note that $m \in \mathcal{M}(|P|, d)$ for every time state $z = (m, u)$.

Definition 4.11 (firing) *Let $\mathcal{D} = (P, T, F, V, m_o, D)$ be a Timed Petri net, $z_1 = (m^{(1)}, u_1)$ a time state in $\mathcal{D}$ and $M \subseteq T$. If M is a maximal step in $m_{.1}^{(1)}$, then M can fire in z_1. After M fires, the net $\mathcal{D}$ is in the time state $z_2 = (m^{(2)}, u_2)$ (notation: $z_1 \xrightarrow{M} z_2$) defined by*

$$m_{i,j}^{(2)} := \begin{cases} m_{i,j}^{(1)} - \sum_{t_s \in M} V(p_i, t_s) + \sum_{\substack{t_s \in M, \\ d_s = 0}} V(t_s, p_i) & \text{if } j = 1 \\ m_{i,j}^{(1)} + \sum_{\substack{t_s \in M, \\ d_s = j - 1}} V(t_s, p_i) & \text{if } j > 1 \end{cases}$$

and

$$u_2(t) := \begin{cases} D(t) & \text{if } t \in M \\ u_1(t) & \text{otherwise} \end{cases}.$$

Definition 4.12 (elapsing of time) *Let $z_1 = (m^{(1)}, u_1)$ be a time state in the Timed Petri net $\mathcal{D}$. Then one time unit can elapse in $\mathcal{D}$ (notation: $z_1 \xrightarrow{1}$), if*

$$\forall t \left((t \in T \wedge u_1(t) = 0) \longrightarrow t^- \not\leq m_{.1}^{(1)} \right).$$

After one time unit has elapsed, $\mathcal{D}$ is in the state $z_2 = (m^{(2)}, u_2)$ (notation: $z_1 \xrightarrow{1} z_2$) defined by

$$m_{i,j}^{(2)} := \begin{cases} m_{i,1}^{(1)} + m_{i,2}^{(1)} & \text{if } j = 1 \\ m_{i,j+1}^{(1)} & \text{if } 2 \leq j \leq d - 1 \\ 0 & \text{if } j = d \end{cases}.$$

We now introduced the notion of a *global step* and a *global reachable time state* (*short: g-reachable*). A global step is a sequence of maximal states complying with certain conditions. The set of all g-reachable time states is a subset of the set of reachable time states of a Timed Petri net.

Definition 4.13 (global step) *Let z be a time state in the Timed Petri net $\mathcal{D}$. A global step in z is the multiset[1] $\mathfrak{G}$ computed by the following procedure:*

1. *$\mathfrak{G} := \emptyset$;*

2. *Let M be a maximal step in z;*

3. **if** *$M \neq \emptyset$* **then** *$\mathfrak{G} := \mathfrak{G} + M$* **else** *stop.*

4. *Let $z \xrightarrow{M} z_1$; Set $z := z_1$;* **goto** *2;*

---

[1]A multiset over a set M denotes a function κ_M from M to the natural numbers. $\kappa_M(x)$ is called the multiplicity of the element x in M.

In $\mathcal{D}_3$ the multiset $\mathfrak{G} = \{t_1, t_2, t_2, t_4\}$ is a global step in z_0. This multiset contains three maximal steps. The set $M_1 = \{t_1, t_2\}$ is a maximal step in z_0. Let $z_0 \xrightarrow{M_1} z_1$. Then $M_2 = \{t_2, t_4\}$ is a maximal step in z_1. Let $z_1 \xrightarrow{M_2} z_2$. Now $M_3 = \emptyset$ is a maximal step in z_2. So we obtain the global step $\mathfrak{G} = M_1 + M_2$.

After firing a maximal step it is generally still possible to fire another maximal step but after firing a global step it is only possible to let time elapse. Therefore, the subsequences of maximal steps of a sequence that are followed by the passing of single time units are global steps, so that the whole sequence alternates between firing global steps and letting single time units pass. So when writing a firing sequence $\sigma = \mathfrak{G}_1 \mathfrak{G}_2 \ldots \mathfrak{G}_n$ of global steps $\mathfrak{G}_i, i \leq i \leq n$, we know that it will result in one of the two sequences $\mathfrak{G}_1 1 \mathfrak{G}_2 1 \ldots 1 \mathfrak{G}_n$ or $\mathfrak{G}_1 1 \mathfrak{G}_2 1 \ldots 1 \mathfrak{G}_n 1$.

Finally, let it be noted that there are Timed Petri nets in which the computation of a global step cannot be completed. The reason is that transitions with duration zero can repeatedly be enabled and then are required to fire infinitely often. To avoid this, one needs to ensure that the net contains no cycles of places and transitions of duration zero. There are however Timed Petri nets containing such cycles where it is not necessary to fire transitions with duration zero infinitely often.

Definition 4.14 (global reachability) *Let $\mathcal{D}$ be a Time Petri net. The time state z^* is globally reachable in $\mathcal{D}$ (short: g-reachable), if there is a sequence $\sigma = \mathfrak{G}_1 \mathfrak{G}_2 \ldots \mathfrak{G}_k$ such that*

$$z^{(0)} \xrightarrow{\mathfrak{G}_1} \hat{z}^{(1)} \xrightarrow{1} \tilde{z}^{(1)} \xrightarrow{\mathfrak{G}_2} \hat{z}^{(2)} \xrightarrow{1} \tilde{z}^{(2)} \ldots \rightarrow z^*.$$

The last change of state can be either the firing of a global step or the elapse of one time unit.

We can analogously define *g-reachable time p-markings*.

In the Timed Petri net $\mathcal{D}_3$ the following time p-marking is reachable but not g-reachable.

$$m = \begin{pmatrix} 1 & 0 & 0 \\ 0 & 0 & 2 \\ 1 & 0 & 0 \end{pmatrix}.$$

Any time state which is reachable but not g-reachable can be reached from some g-reachable time state by firing maximal steps.

Definition 4.15 (time-extended incidence matrix) *Let* $\mathcal{D} = (P, T, F, V, m_0, D)$ *be a Timed Petri net. The time-extended incidence matrix of* $\mathcal{D}$ *is the matrix* $C \in \mathcal{M}(|P|, d \cdot |T|)$, $C := (C^{(1)}, C^{(2)}, \ldots, C^{(|T|)})$ *with* $C^{(k)} \in \mathcal{M}(|P|, d)$, $k \in \{1, \ldots, |T|\}$ *and submatrices* $C^{(k)} = (c_{i,j}^{(k)})$ *defined as follows:*

$$
c_{i,j}^{(k)} := \begin{cases}
V(t_k, p_i) - V(p_i, t_k) & \text{if } d_k = 0, \, j = 1 \\
-V(p_i, t_k) & \text{if } d_k > 0, \, j = 1 \\
V(t_k, p_i) & \text{if } d_k > 0, \, 0 < j - 1 = d_k \\
0 & \text{otherwise}
\end{cases}
$$

Example 4.16 *The time-extended incidence matrix of the Timed Petri net* $\mathcal{D}_3$ *(Fig. 4.7) is*

$$
C = \begin{pmatrix}
-2 & 0 & 0 & -1 & 0 & 0 & 0 & 0 & 3 & 0 & 0 & 2 \\
0 & 0 & 1 & 0 & 0 & 0 & -1 & 0 & 0 & 0 & 0 & 0 \\
0 & 0 & 0 & 1 & 0 & 0 & 0 & 0 & 0 & -1 & 0 & 0
\end{pmatrix}.
$$

Definition 4.17 (bag matrix) *Let* $\mathcal{D} = (P, T, F, V, m_0, D)$ *be a Timed Petri net and let* $\mathfrak{G}$ *be a global step in a reachable time state of* $\mathcal{D}$ *and let the multiplicity* $\kappa_{\mathfrak{G}}(t_i)$ *of* t_i *in* $\mathfrak{G}$ *be denoted by* $\kappa_{\mathfrak{G}_i}$. *The bag matrix of* $\mathfrak{G}$ *is the following Matrix* $G \in \mathcal{M}(d \cdot |T|, d)$:

$$
G = (g_{i,j})_{\substack{i=1\ldots d \cdot |T| \\ j=1\ldots d}} = \begin{pmatrix}
G_{(1)} \\
G_{(2)} \\
\vdots \\
G_{(|T|)}
\end{pmatrix} \quad \text{with } G_{(s)} = \kappa_{\mathcal{G}_s} \cdot E_d.
$$

If the transition t_s does not belong to the global step $\mathfrak{G}$, then the submatrix $G_{(s)}$ is obviously a zero matrix.

Example 4.18 *Let us consider the global step* $\mathfrak{G} = \{t_1, t_2, t_2, t_4\} = \{t_1, t_2\} + \{t_2, t_4\}$ *in the time state* $z^{(0)}$ *in* $\mathcal{D}_3$. *The bag matrix* G *of* $\mathfrak{G}$ *is as follows:*

$$G = \begin{pmatrix} G_{(1)} \\ G_{(2)} \\ G_{(3)} \\ G_{(4)} \end{pmatrix} \quad with \quad G_{(1)} = \begin{pmatrix} 1 & 0 & 0 \\ 0 & 1 & 0 \\ 0 & 0 & 1 \end{pmatrix}, \ G_{(2)} = \begin{pmatrix} 2 & 0 & 0 \\ 0 & 2 & 0 \\ 0 & 0 & 2 \end{pmatrix},$$

$$G_{(3)} = \begin{pmatrix} 0 & 0 & 0 \\ 0 & 0 & 0 \\ 0 & 0 & 0 \end{pmatrix}, \ G_{(4)} = \begin{pmatrix} 1 & 0 & 0 \\ 0 & 1 & 0 \\ 0 & 0 & 1 \end{pmatrix}, \ i.e., \quad G = \begin{pmatrix} 1 & 0 & 0 \\ 0 & 1 & 0 \\ 0 & 0 & 1 \\ 2 & 0 & 0 \\ 0 & 2 & 0 \\ 0 & 0 & 2 \\ 0 & 0 & 0 \\ 0 & 0 & 0 \\ 0 & 0 & 0 \\ 1 & 0 & 0 \\ 0 & 1 & 0 \\ 0 & 0 & 1 \end{pmatrix}.$$

Definition 4.19 (progress matrix) *Let $\mathcal{D}$ be a Timed Petri net with time dimension d. The progress matrix of $\mathcal{D}$ is the matrix $R \in \mathcal{M}(d,d)$ with*

$$r_{i,j} := \begin{cases} 1 & if \ (i = j = 1) \ or \ (i = j + 1) \\ 0 & otherwise \end{cases} .$$

Example 4.20 *The progress matrix R of $\mathcal{D}_3$ is $R = \begin{pmatrix} 1 & 0 & 0 \\ 1 & 0 & 0 \\ 0 & 1 & 0 \end{pmatrix}$.*

Definition 4.21 (Parikh matrix) *Let $\mathcal{D}$ be a Timed Petri net and $\sigma = \mathfrak{G}_1 \ldots \mathfrak{G}_n$ a firing sequence in $\mathcal{D}$. Furthermore let $G^{(i)}$ be the bag matrix of the global step $\mathfrak{G}_i$ for every $i = 1, \ldots, n$. The Parikh matrix of σ is the matrix $\Psi \in \mathcal{M}(d \cdot |T|, d)$ with*

$$\Psi_\sigma := \sum_{i=1}^{n} G^{(i)} \cdot R^{n-i}.$$

Now we can finally introduce the notion of *state equations* for Timed Petri nets.

Theorem 4.22 (state equations for Timed Petri nets) *Let $\mathcal{D}$ be a Timed Petri net with progress matrix R and the initial time state $z^{(0)} = (m^{(0)}, u_0)$ and let $\sigma = \mathfrak{G}_1 \ldots \mathfrak{G}_n$ be a firing sequence in $\mathcal{D}$ with $z^{(0)} \xrightarrow{\mathfrak{G}_1} \hat{z}^{(1)} \xrightarrow{1} \tilde{z}^{(1)} \xrightarrow{\mathfrak{G}_2} \hat{z}^{(2)} \xrightarrow{1} \ldots \xrightarrow{\mathfrak{G}_n} \hat{z}^{(n)}$. Then it holds that:*

$$\hat{m}^{(n)} = m^{(0)} \cdot R^{n-1} + C \cdot \Psi_\sigma, \tag{2}$$

where $\hat{z}^{(i)} = (\hat{m}^{(i)}, u_i)$ and $\tilde{z}^{(i)} = (\tilde{m}^{(i)}, u_i)$ for every $i \in \{1, \ldots, n\}$.

The equality (2) is called the *first state equation* for the firing sequence σ in $\mathcal{D}$.

For the proof of Theorem 4.22 we need the next two lemmas:

Lemma 4.23 *Let $\mathcal{D}$ be a Timed Petri net, R its progress matrix, $z^{(1)} = (m^{(1)}, u_1)$ a reachable time state in $\mathcal{D}$ and $z^{(1)} \xrightarrow{1} z^{(2)}$ with $z^{(2)} = (m^{(2)}, u_2)$. Then it holds that:*
$$m^{(2)} = m^{(1)} \cdot R.$$

Proof:

Let us consider the matrix $Q := m_{i,j}^{(1)} \cdot R$ with $Q = (q_{i,j})$ for $1 \leq i \leq |P|$ and $1 \leq j \leq d$. In order to prove the lemma it is now sufficient to show that $q_{i,j} = m_{i,j}^{(2)}$.

Case 1: $j = 1$

Hence, it holds that: $q_{i,1} = \sum_{s=1}^{d} m_{i,s}^{(1)} \cdot r_{s,1} = m_{i,1}^{(1)} \cdot 1 + m_{i,2}^{(1)} \cdot 1 \underset{Def.4.12}{=} m_{i,1}^{(2)}$

Case 2: $d > j \geq 2$

Hence, it holds that: $q_{i,j} = \sum_{s=1}^{d} m_{i,s}^{(1)} \cdot r_{s,j} = m_{i,j+1}^{(1)} \cdot 1 \underset{Def.4.12}{=} m_{i,j}^{(2)}$

Case 3: $j = d$

Hence, it holds that: $q_{i,j} = \sum_{s=1}^{d} m_{i,s}^{(1)} \cdot r_{s,j} = 0 = m_{i,j}^{(2)}$. $\square$

Lemma 4.24 *Let $\mathcal{D}$ be a Timed Petri net and $z^{(1)} = (m^{(1)}, u_1)$ a reachable time state in $\mathcal{D}$. Furthermore let $z^{(1)} \xrightarrow{\mathfrak{G}} z^{(2)}$ with $z^{(2)} = (m^{(2)}, u_2)$ and $\mathfrak{G} = \{\kappa_{\mathcal{G}_{i_1}} \cdot t_{i_1}, \ \kappa_{\mathcal{G}_{i_2}} \cdot t_{i_2}, \ldots, \ \kappa_{\mathcal{G}_{i_q}} \cdot t_{i_q}\}$. Then it holds that:*

$$m^{(2)} = m^{(1)} + C \cdot G.$$

Proof:

We compute $m_{i,j}^{(2)}$.

Case 1: $j = 1$

By Definitions 4.11 and 4.13 it follows that:

$$m_{i,1}^{(2)} = m_{i,1}^{(1)} - \sum_{t_s \in \mathcal{G}} \left(V(p_i, t_s) \cdot \kappa_{\mathcal{G}_s} \right) + \sum_{\substack{d_s = j-1 = 0 \\ t_s \in \mathcal{G}}} \left(V(t_s, p_i) \cdot \kappa_{\mathcal{G}_s} \right).$$

Hence, it is sufficient to show that

$$\sum_{l=1}^{d \cdot |T|} c_{i,l} g_{l,1} = - \sum_{t_s \in \mathcal{G}} \left(V(p_i, t_s) \cdot \kappa_{\mathcal{G}_s} \right) + \sum_{\substack{d_s = j-1 = 0 \\ t_s \in \mathcal{G}}} \left(V(t_s, p_i) \cdot \kappa_{\mathcal{G}_s} \right).$$

We consider the bag matrix G. The first column $G_{.1}$ of G is the vector $G_{.1} = (g^{(1)}, g^{(2)}, \ldots, g^{(|T|)})^T$, where $g^{(k)}$ is a d-dimensional vector with

$$g^{(k)} = (g_1^{(k)}, \ldots, g_d^{(k)})^T \text{ for } k = 1, \ldots, |T| \text{ and } g_j^{(k)} = \begin{cases} \kappa_{\mathfrak{G}_k} & \text{if } j = 1 \\ 0 & \text{otherwise} \end{cases}.$$

So, if $t_k \in \mathfrak{G}$ then $g^{(k)}$ is the d-dimensional vector $(\kappa_{\mathcal{G}_k}, 0, \ldots, 0)^T$ and otherwise $g^{(k)}$ is the d-dimensional zero vector.

Thus, it holds that:

$$
\begin{aligned}
\sum_{l=1}^{d \cdot |T|} c_{i,l} \cdot g_{l,1} &= \sum_{k=1}^{|T|} \sum_{l=1}^{d} c_{i,l}^{(k)} \cdot g_l^{(k)} \\
&= \sum_{k=1}^{|T|} \left(c_{i,1}^{(k)} \cdot \kappa_{\mathcal{G}_k} \right) \qquad\qquad \text{by Def. 4.15, for } j = 1 \\
&= - \sum_{t_k \in \mathcal{G}} \left(V(p_i, t_k) \cdot \kappa_{\mathcal{G}_k} \right) + \sum_{\substack{d_k = 0 \\ t_k \in \mathcal{G}}} \left(V(t_k, p_i) \cdot \kappa_{\mathcal{G}_k} \right).
\end{aligned}
$$

Case 2: $j \geq 2$

We now have to show that $\displaystyle\sum_{l=1}^{d \cdot |T|} c_{i,l} \cdot g_{l,j} = \sum_{d_s=j-1} \left(V(t_s, p_i) \cdot \kappa_{\mathcal{G}_s} \right)$.

Let us consider the j-th column of the bag matrix G, which is the vector $G_{\cdot j}$. It holds that:

$$G_{\cdot j} = (g^{(1)}, g^{(2)}, \ldots, g^{(|T|)})^T \quad \text{with} \quad \overset{|T|}{\underset{1}{\forall}} k \quad g^{(k)} = (g_1^{(k)}, \ldots, g_d^{(k)})^T$$

$$\text{and} \quad g_l^{(k)} = \begin{cases} \kappa_{\mathcal{G}_k} & \text{if } l = j \\ 0 & \text{otherwise,} \end{cases}$$

i.e., $g^{(k)} = \underbrace{(0, \ldots, 0)^T}_{d}$, if $t_k \notin \mathcal{G}$

and $g^{(k)} = \underbrace{(0, \ldots, \underset{\uparrow j.}{\kappa_{\mathcal{G}_k}}, 0, \ldots, 0)^T}_{d}$, if $t_k \in \mathcal{G}$.

This yields

$$\sum_{l=1}^{d \cdot |T|} c_{i,l} \cdot g_{l,j} = \sum_{k=1}^{|T|} \left(c_{i,j}^{(k)} \cdot \kappa_{\mathcal{G}_k} \right) \qquad \text{by Def. 4.15, for } j \geq 2$$

$$= \sum_{\substack{d_k=j-1 \\ t_k \in \mathcal{G}}} \left(V(t_k, p_i) \cdot \kappa_{\mathcal{G}_k} \right).$$

$\square$

Proof of Theorem 4.22:

We will do the proof by induction on n:

Basis: We have to show that for $m^{(0)} \xrightarrow{\mathcal{G}_1} m^{(1)}$ it holds that

$$m^{(1)} = m^{(0)} \cdot R^0 + C \cdot \Psi_\sigma = m^{(0)} + C \cdot \Psi_\sigma, \text{ with}$$

$$\Psi_\sigma = \sum_{i=1}^{1} G^{(i)} \cdot R^{n-i} = G^{(1)} \cdot R^0 = G^{(1)}.$$

Hence, we have to prove that $m^{(1)} = m^{(0)} + C \cdot G^{(1)}$. This is true because of Lemma 4.24.

Step: We consider the firing sequence $\sigma' = \sigma \mathfrak{G}_{n+1}$ with

$$z^{(0)} \xrightarrow{\sigma} \hat{z}^{(n)} \xrightarrow{1} \tilde{z}^{(n)} \xrightarrow{\mathfrak{G}_{n+1}} \hat{z}^{(n+1)}.$$

We have to show that the following equation holds:

$$\hat{m}^{(n+1)} = m^{(0)} \cdot R^n + C \cdot \Psi_{\sigma'} \quad \text{and} \quad \Psi_{\sigma'} = \sum_{i=1}^{n+1} G^{(i)} \cdot R^{n+1-i}.$$

According to the induction hypothesis it holds that:

$$\hat{m}^{(n)} = m^{(0)} \cdot R^{n-1} + C \cdot \Psi_\sigma = m^{(0)} \cdot R^{n-1} + C \cdot \left(\sum_{i=1}^{n} G^{(i)} \cdot R^{n-i} \right). \tag{3}$$

Furthermore, because of Lemma 4.23 and Lemma 4.24 it holds that

$$\tilde{m}^{(n)} = \hat{m}^{(n)} \cdot R \tag{4}$$

and

$$\hat{m}^{(n+1)} = \tilde{m}^{(n)} + C \cdot G^{(n+1)}. \tag{5}$$

Hence, (3), (4) and (5) immediately yield:

$$\hat{m}^{(n+1)} \underset{(5)}{=} \tilde{m}^{(n)} + C \cdot G^{(n+1)} \underset{(4)}{=} \hat{m}^{(n)} \cdot R + C \cdot G^{(n+1)}$$

$$\underset{(3)}{=} \left(m^{(0)} \cdot R^{n-1} + C \cdot \left(\sum_{i=1}^{n} G^{(i)} R^{n-i} \right) \right) \cdot R + C \cdot G^{(n+1)}$$

$$= m^{(0)} \cdot R^n + C \cdot \left(\sum_{i=1}^{n} G^{(i)} R^{n+1-i} \right) + C \cdot G^{(n+1)} \cdot \underbrace{E_d}_{=R^0}$$

$$= m^{(0)} \cdot R^n + C \cdot \left(\sum_{i=1}^{n} G^{(i)} R^{n+1-i} + G^{(n+1)} \cdot R^{(n+1)-(n+1)} \right)$$

$$= m^{(0)} \cdot R^n + C \cdot \left(\sum_{i=1}^{n+1} G^{(i)} R^{n+1-i} \right)$$

$$= m^{(0)} R^n + C \cdot \Psi_{\sigma'}.$$

□

Corollary 4.25 *Let $\mathcal{D}$ be a Timed Petri net with progress matrix R and initial time state $z^{(0)}$ and let $\sigma = \mathfrak{G}_1 \ldots \mathfrak{G}_n$ be a firing sequence in $\mathcal{D}$ with*

$z^{(0)} \xrightarrow{\mathfrak{G}_1} \hat{z}^{(1)} \xrightarrow{1} \tilde{z}^{(1)} \xrightarrow{\mathfrak{G}_2} \hat{z}^{(2)} \xrightarrow{1} \ldots \xrightarrow{1} \tilde{z}^{(n)}$. *Then it holds that:*

$$\tilde{m}^{(n)} = m^{(0)} \cdot R^n + C \cdot \Psi_\sigma R. \tag{6}$$

The equality (6) is called the *second state equation* for the firing sequence σ in $\mathcal{D}$.

Let $\sigma = \mathfrak{G}_1\mathfrak{G}_2 \ldots \mathfrak{G}_n$ be a firing sequence in the Timed Petri net $\mathcal{D}$ and let $z^{(1)} \xrightarrow{\sigma} z^{(2)}$. If all global steps in σ are empty, then we often write $z^{(1)} \xrightarrow{n} z^{(2)}$ or $z^{(1)} \xrightarrow{n-1} z^{(2)}$ (depending on the last state change).

Corollary 4.26 *Let $z^{(1)} \xrightarrow{\tau} z^{(2)}$ for some $\tau \in \mathbb{N}^+$ in the Timed Petri net $\mathcal{D}$. Then it holds that:*

$$\forall j \left((1 \leq j \leq |P|) \to \sum_{i=1}^{d} m_{j,i}^{(1)} = \sum_{i=1}^{d} m_{j,i}^{(2)} \right). \tag{7}$$

Remark 4.27 *Let $z^{(1)} \xrightarrow{M} z^{(2)}$ with M a maximal step in $z^{(1)}$ in the Timed Petri net $\mathcal{D}$. Then it holds that*

$$\forall t \left((t \in (T \setminus M) \wedge u_2(t) = 0) \longrightarrow m_{.1}^{(1)} - \sum_{\hat{t} \in M} \hat{t}^- \not\geq t^- \right). \tag{8}$$

Remark 4.28 *Let $z^{(1)} \xrightarrow{\mathfrak{G}} z^{(2)}$ in the Timed Petri net $\mathcal{D}$. Then it holds that:*

$$\forall t \left((t \in T \wedge u_2(t) = 0) \longrightarrow m_{.1}^{(2)} \not\geq t^- \right). \tag{9}$$

Theorem 4.22, Corollary 4.26 and Remarks 4.27 and 4.28 give necessary conditions for the reachability of time p-markings. Therefore, they can be used to prove non-reachability of specific time states. Such conditions are important because the reachability of a time p-marking is in general not decidable.

The following Example 4.29 illustrates the use of state equations in proving non-reachability of p-markings:

Example 4.29 *Using Theorem 4.22 and Corollary 4.26 we will show that the time p-marking*

$$m^* = \begin{pmatrix} 1 & 0 \\ 0 & 1 \end{pmatrix}$$

is not reachable in $\mathcal{D}_4$ (see Fig. 4.8).

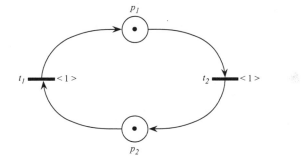

Figure 4.8: The Timed Petri net $\mathcal{D}_4$

Assuming m^* is reachable in $\mathcal{D}_4$, there exists a firing sequence $\sigma = \mathfrak{G}_1 \mathfrak{G}_2 \ldots \mathfrak{G}_n$ such that $z^{(0)} \xrightarrow{\sigma} z^{(n)}$ and $z^{(n)} = z^*$. Then equality (2) or (6) holds for $m^* = m^{(n)}$.

The time-extended incidence matrix C, the progress matrix R and $R^n, n \geq 1$ of $\mathcal{D}_4$ are:

$$C = \begin{pmatrix} 0 & 1 & -1 & 0 \\ -1 & 0 & 0 & 1 \end{pmatrix}, R = \begin{pmatrix} 1 & 0 \\ 1 & 0 \end{pmatrix}, R^n = \begin{pmatrix} 1 & 0 \\ 1 & 0 \end{pmatrix}.$$

Let x_j^i be the number of appearances of the transition t_j in $\mathfrak{G}_i$. The bag matrix G_i of the global step $\mathfrak{G}_i$ $(1 \leq i \leq n)$ is then as follows:

$$G^{(i)} = \begin{pmatrix} x_1^i & 0 \\ 0 & x_1^i \\ x_2^i & 0 \\ 0 & x_2^i \end{pmatrix}.$$

Therefore, the Parikh matrix Ψ_σ is:

$$\Psi_\sigma = \begin{pmatrix} x_1^1 + \ldots + x_1^n & 0 \\ x_1^1 + \ldots + x_1^{n-1} & x_1^n \\ x_2^1 + \ldots + x_2^n & 0 \\ x_2^1 + \ldots + x_2^{n-1} & x_2^n \end{pmatrix}.$$

Case 1: *Equality (2) holds for m^*. Then it holds that*

$$C \cdot \Psi_\sigma = \begin{pmatrix} 0 & 1 & -1 & 0 \\ -1 & 0 & 0 & 1 \end{pmatrix} \cdot \begin{pmatrix} x_1^1 + \ldots + x_1^n & 0 \\ x_1^1 + \ldots + x_1^{n-1} & x_1^n \\ x_2^1 + \ldots + x_2^n & 0 \\ x_2^1 + \ldots + x_2^{n-1} & x_2^n \end{pmatrix}.$$

This leads to

$$m^* = \begin{pmatrix} 1 & 0 \\ 0 & 1 \end{pmatrix} = \begin{pmatrix} 1 & 0 \\ 1 & 0 \end{pmatrix} \cdot \begin{pmatrix} 1 & 0 \\ 1 & 0 \end{pmatrix} + \begin{pmatrix} \alpha - \beta - x_2^n & x_1^n \\ -\alpha - x_1^n + \beta & x_2^n \end{pmatrix},$$

i.e.,

$$\begin{pmatrix} 0 & 0 \\ -1 & 1 \end{pmatrix} = \begin{pmatrix} \alpha - \beta - x_2^n & x_1^n \\ -\alpha - x_1^n + \beta & x_2^n \end{pmatrix}, \tag{10}$$

with $\alpha := x_1^1 + \ldots + x_1^{n-1}$ and $\beta := x_2^1 + \ldots + x_2^{n-1}$ for $n \geq 2$.
Now (10) yields

$$x_1^n = 0, \tag{11}$$
$$\alpha = \beta + x_2^n, \; x_2^n = 1 \; and \; -1 = -\alpha - x_1^n + \beta.$$

Thus, it follows that

$$x_1^1 \neq x_2^1 \tag{12}$$

in contradiction to $\mathfrak{G}_1 = \{t_1, t_2\}$. Therefore we have ruled out this case.

Case 2: *Equality (6) holds for m^*.*

Then, it follows that:

$$m^* = \begin{pmatrix} 1 & 0 \\ 0 & 1 \end{pmatrix} = \begin{pmatrix} 1 & 0 \\ 1 & 0 \end{pmatrix} + \begin{pmatrix} \alpha - \beta - x_2^n & x_1^n \\ -\alpha - x_1^{n-1} + \beta & x_2^n \end{pmatrix} \cdot \begin{pmatrix} 1 & 0 \\ 1 & 0 \end{pmatrix}.$$

Hence

$$\begin{pmatrix} 0 & 0 \\ -1 & 1 \end{pmatrix} = \begin{pmatrix} \alpha + x_1^n - \beta - x_2^n & 0 \\ -\alpha - x_1^{n-1} + \beta + x_2^n & 0 \end{pmatrix},$$

and therefore $1 = 0$ which is obviously again a contradiction. Therefore, this case cannot hold either.

Thus, we have shown that the assumption of m^* being reachable in $\mathcal{D}_4$ inevitably leads to contradictions. Hence, m^* is not reachable in $\mathcal{D}_4$.

4.5   Variations of Timed Petri Nets

Petri nets are suitable for modeling systems on different levels of abstraction. It is the desire of every user to describe a system adequately, exactly, and in a natural way. This has led to the introduction of various extensions of Timed Petri nets. But the more extensions are defined, the more difficult and limited the analysis becomes. In the next two subsections we present two useful extensions of Timed Petri nets.

4.5.1   Timed Petri Nets with Priorities

One enrichment of the specification possibilities of Timed Petri Nets is the addition of priorities. This extension became necessary in order to support the specification of priority-based schedules as used in real-time systems. An example of the use of a specification using Timed Petri Nets with priorities is the verification of the architecture "Message Scheduled System" (MSS) that targets control systems for automation and transportation. More on the specification and the analysis of this architecture can be found in [RWPZ02] and [Ric06].

Definition 4.30 *A Timed Petri net with priorities (P-Timed Petri net) is a pair* $\mathcal{PD} = (\mathcal{D}, \xi)$ *such that*

   1. $\mathcal{D}$ is a Timed Petri net with T as its set of transitions and

   2. $\xi : T \longrightarrow \mathbb{N}$.

The function ξ is called the *priority function* of $\mathcal{PD}$. In the graphic representation of a P-Timed Petri net the value $\xi(t)$ is given in curly brackets. The value $D(t)$ is given in angle brackets.

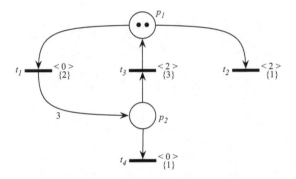

Figure 4.9:   The P-Timed Petri net $\mathcal{PD}_1$

All the notions defined for Timed Petri nets can be adapted to P-Timed Petri nets. In deciding which transitions to use for a maximal step in a P-Timed Petri net, transitions with higher priorities are always preferred over transitions with lower priorities.

Definition 4.31 (prioritized maximal step) *Let $z = (m, u)$ be a time state in the P-Timed Petri net $\mathcal{PD}$ with set of transitions T. Then $M \subseteq T$ is a prioritized maximal-step in z, if*

   (1) $\forall t \, (\, t \in M \longrightarrow u(t) = 0 \,)$,

(2) $\sum\limits_{t\in M} t^- \leq m_{.1},$

(3) $\forall \hat{t} \left((\hat{t}^- \leq m_{.1} \wedge u(\hat{t}) = 0 \wedge \hat{t} \notin M) \longrightarrow \sum\limits_{\substack{t\in M, \\ \xi(t)\geq\xi(\hat{t})}} t^- + \hat{t}^- \nleq m_{.1} \right),$

(4) $\neg\exists M^* \left((M^* \supsetneq M) \wedge (M \text{ satisfies } (1) - (4)) \right).$

We note that it is possible to obtain a prioritized maximal step containing a transition with a lower priority than some enabled transition which does not belong to the step: Consider the net $\mathcal{PD}_2$ in Figure 4.10. The only prioritized maximal step in the initial time state is $M = \{t_1, t_3\}$. Transition t_2 is not in M and t_3 is in M even though t_2 has higher priority than t_3. The reason is that t_2 is in a (dynamic) conflict with transition t_1 which has higher priority than t_2.

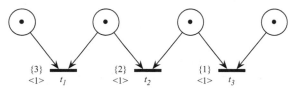

Figure 4.10: The P-Timed Petri net $\mathcal{PD}_2$

The firing rule for P-Timed Petri nets is *immediate firing in maximal step*. All formal definitions for P-Timed Petri nets can be found in [WPZR04]. They are similar to the definitions in Section 4.4. For P-Timed Petri nets we can give an additional sufficient condition for non-reachability which makes use of the priorities. Using this condition and the time-dependent state equations for the net $\mathcal{PD}_1$ we can for instance prove that the time p-marking

$$m^* = \begin{pmatrix} 0 & 0 & 0 \\ 0 & 0 & 0 \end{pmatrix}$$

is not reachable in $\mathcal{PD}_1$. The approach is similar to the one in Example 4.29.

Finally, we note that reachability graphs for P-Timed Petri nets can be defined without difficulty and hence graph-theoretical algorithms can be applied in the analysis of these nets.

4.5.2   Timed Petri Nets with Variable Durations

In the real world it is often difficult or even impossible to determine an exact duration for any event but it might be possible to estimate lower and upper bounds for the duration of the event. To model such circumstances with time-dependent Petri nets, every transition is associated with an interval determining the range of possible durations of this transition. Such nets are called Timed Petri nets with variable durations.

Definition 4.32 (Timed Petri nets with variable durations) *A Timed Petri net with variable durations (short: U-Timed Petri net) is a pair* $\mathcal{UD} =$ $(\mathcal{N}, U)$ *such that*

1. $\mathcal{N}$ *is a Petri net, T is its set of transitions and*

2. $U : T \longrightarrow \mathbb{Q}^+ \times \mathbb{Q}^+$ *with $U(t) = (a_t, b_t)$ implies $a_t \le b_t$ for every $t \in T$.*

The duration of the firing process of a transition is not fixed, it varies within the interval associated with the transition by the function U. The firing mode is immediate firing of maximal steps. A formal introduction of U-Timed Petri nets and of state equations for these nets can be found in [PZP12]. Fig. 4.11 shows a U-Timed Petri net.

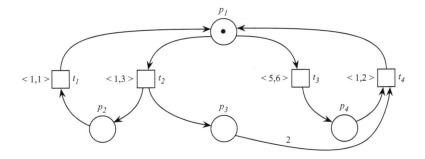

Figure 4.11:   The U-Timed Petri net $\mathcal{UD}_1$

One way to analyze a U-Timed Petri net is to transform it into a Time Petri net and then analyze the obtained net. The transformation is similar to the transformation of Timed Petri nets into Time Petri nets in Section 4.3, cf. Fig. 4.12.

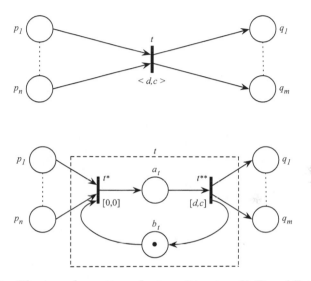

Figure 4.12: The transformation of a transition in a U-Timed Petri net into a Time Petri net

For U-Timed Petri nets containing transitions with zero as a lower or upper bound there is no such straightforward transformation, but individual nets can be transformed similarly to Timed Petri nets with zero durations, cf. Fig. 4.5.

These nets are mostly used in the modeling and verification of the dependability of technical systems (dependability engineering). We used them for software verification in [HPZ97].

4.6   Bibliographical Notes

Unlike in classical and Time Petri nets, in Timed Petri nets it is completely determined when in time the transitions fire, so that no real concurrency exists in these nets. Thus, every Timed Petri net has an implicit global clock. But we still do not need a global clock for the analysis, it is sufficient to consider time locally at each transition. The reason is that time passes at the same speed for all firing transitions. This does not mean that time always passes at equal speed.

Apart from Ramchandani who first introduced Timed Petri nets, Sifakis in [Sif80] and Starke in [Sta90] have studied these nets. W. M. Zuberek extensively studied Timed Petri nets and a wide area of their applications and made huge theoretical and practical contributions in this area. An algebraic method for the quantitative analysis of conflict-free Timed Petri nets (Timed Petri nets whose skeletons are marked graphs) is given in [Wan98]. The method, however, cannot be generalized to arbitrary Timed Petri nets.

4.7 Exercises

Exercise 4.1

Let $\mathcal{D} = (P, T, V, m_0, D)$ be a Timed Petri net and $\mathcal{Z}_\mathcal{D}$ the Time Petri net obtained from $\mathcal{D}$ using the local transformation described in Section 4.3 (cf. Fig. 4.4). For each p-marking m in $\mathcal{D}$ we let the set $M_m \subseteq R_{\mathcal{Z}_\mathcal{D}}$ be defined as follows:

$$M_m := \{\tilde{m} \mid \forall p \, (p \in P \to \tilde{m}(p) := \begin{cases} m(p) & \text{if } p \in P \\ 0 \text{ or } 1 & \text{otherwise} \end{cases})\}.$$

Show that if m is reachable in $\mathcal{D}$, then at least one element $\tilde{m}$ of M_m is a reachable p-marking in $\mathcal{Z}_\mathcal{D}$.

Remark: *The set M_m is nonempty and finite for every reachable p-marking in $\mathcal{D}$.*

Exercise 4.2

Define the notions of "maximal step" and "global step" for U-Timed Petri nets with the additional condition that the duration of any transition should be a natural number. Is it possible to generalize these definitions so that they allow rational durations?

Exercise 4.3

Give a definition of a time-dependent state equation for a sequence of global steps and elapses of single time units in U-Timed Petri nets where the durations of all transitions are natural numbers.

Exercise 4.4

(a) Consider the following Timed Petri net:

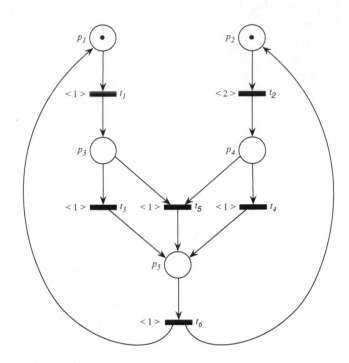

Give a timeless Petri net such that the two nets have isomorphic reachability graphs.

(b) Let $\mathcal{D} = (P, T, V, m_0, D)$ be an arbitrary Timed Petri net. Give a local transformation which transforms $\mathcal{D}$ into a timeless Petri net $\mathcal{N}_D$ such that the reachability graphs of the two nets are isomorphic.

Hint: *Note that Timed Petri nets are Turing equivalent but classic Petri nets are not.*

Chapter 5

Petri Nets with Time Windows
(Petri Nets with Retention Time on Places)

The last time extension of Petri nets introduced in this book are Petri nets with retention time on places and we show ways to analyze them. They were first devised as a formal specification technique for designing complex automaton systems. Most important for this application are the possibilities for system diagnostics, namely error detection, localization, evaluation, recognition and reaction. It is important to minimize the total time needed for the internal diagnostic process of the system; considering only the causal relationships between events is therefore not sufficient.

In order to model event-driven systems in a natural way by Petri nets we would like to be able to specify delays and safety distances of time between events. That is why we extend classic Petri nets by *retention times* for tokens on the places. The time assigned to a place can be interpreted in different ways, as minimum or maximum delay, or a combination of both, as exact retention time or periods of validity, etc.

We consider here a time extension where for each place an interval specifies the minimum and maximum retention time for tokens reaching this place. When a token arrives in a place, "its time" starts to run. If the token remains on the place until after its maximum retention time has passed, we can either define that it may not be used to fire a transition any more at all and simply remains on this place or we can reset its time to zero. The minimum retention time can then be viewed as a kind of preparation time required before the token is ready for use and the token only stays ready for a limited amount of time, namely until its maximum retention time has passed, then it needs to be prepared again and can be used for firing when it has reached its minimum retention time. This second version is the one we consider here.

The minimum and maximum retention time of each place can be understood as a time window for each token on the place. The window is closed until the considered token has reached the minimum retention time. Then the window opens and the token can leave the place but only until the maximum retention time is reached, when the window closes. Therefore these nets are also sometimes called Petri nets with time windows.

5.1   Definitions

Definition 5.1 (Petri nets with retention time on places) *A Petri net with retention time on places or with time windows (short: tw-Petri net), is a pair* $\mathcal{P} = (\mathcal{N}, I)$ *such that*

1. $\mathcal{N}$ *is a classic Petri net,* P *its set of places and*

2. $I : P \longrightarrow \mathbb{Q}_0^+ \times (\mathbb{Q}_0^+ \cup \{\infty\})$ *with* $I(p) = (l_p, u_p)$ *where* $l_p \leq u_p$ *for all* $p \in P.$

The classic Petri net $\mathcal{N}$ is called the skeleton of $\mathcal{P}$ and is denoted by $S(\mathcal{P})$. Without loss of generality we consider the codomain of I to be $\mathbb{N}_0^+ \times (\mathbb{N}_0^+ \cup \{\infty\})$. Fig. 5.1 illustrates a Petri net with retention time on places.

Tokens can obviously arrive on a place at different moments. Therefore each token needs its own clock. This can be formally described in an unexpectedly simple and elegant way by using words over the real numbers. Each token in a place is represented by a (real) number (modulo[1] the upper bound of the place) representing the time this token has spent on the place. Thus, a place with three tokens is marked with a word consisting of three numbers. Within one such word, the numbers will be ordered numerically, decreasing from left to right. An unmarked place is marked with the empty word. Such a marking will be called a *time-marking*.

---

[1]The *modulo* notion will be slightly modified, cf. Definition 5.9

Example 5.2

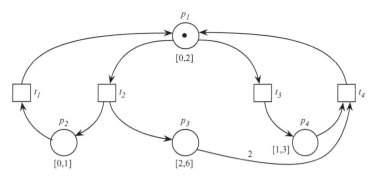

Figure 5.1: $\mathcal{P}_1$ is a Petri net with retention time on places

Definition 5.3 (time-marking) *Let $\mathcal{P}$ be a tw-Petri net with set of places P. A time-marking in $\mathcal{P}$ is a total function $M : P \longrightarrow \left(\mathbb{R}_0^+\right)^*$.*

Definition 5.4 (integer time-marking) *A time-marking M in a tw-Petri net with a set of places P is an integer time-marking, if it holds that $M : P \longrightarrow \mathbb{N}^*$.*

Notation: We write $m_M := (\ell(M(p_1)), \ell(M(p_2)), \ldots, \ell(M(p_{|P|})))$ to denote the marking in $S(\mathcal{P})$ corresponding to the time-marking M in a tw-Petri net $\mathcal{P}$.

Definition 5.5 (initial time-marking) *Let $\mathcal{P}$ be a tw-Petri net and m_0 the initial marking of $S(\mathcal{P})$. The initial time-marking M_0 of $\mathcal{P}$ is defined by*

$$M_0(p) := \begin{cases} \varepsilon & \text{if } m_0(p) = 0 \\ 0^{m_0(p)} & \text{otherwise} \end{cases}.$$

Note that it always holds that: $m_{M_0} = m_0$.

The initial time-marking M_0 of $\mathcal{P}_1$ (cf. Fig. 5.1) is $M_0 = (0, \varepsilon, \varepsilon, \varepsilon)$.

As usual for time-dependent Petri nets, state change can occur through firing of transitions or through elapsing of time. Tw-Petri nets fire in single firing mode. Thus, we first define under which circumstances a transition is *ready to fire* and then introduce state change through firing of transitions.

Definition 5.6 (ready to fire) *Let $\mathcal{P}$ be a tw-Petri net with set of places P, t a transition in $\mathcal{P}$ and M a time-marking in $\mathcal{P}$ with $M(p) = a_1^p a_2^p \ldots a_n^p$ for each place $p \in P$. Then t is ready to fire in M, if*

    1. $t^- \leq m_M$,

    2. $\forall p(p \in {}^\bullet t \longrightarrow \forall j(j \in \{1, \ldots, t^-(p)\} \longrightarrow l_p \leq a_j^p \leq u_p))$.

Definition 5.7 (firing) *Let $\mathcal{P}$ be a tw-Petri net, T its set of transitions and M a time-marking in $\mathcal{P}$. A transition $t \in T$ can fire in M if t is ready to fire in M (notation: $M \xrightarrow{t}$). After the firing of t the time-marking M changes into the time-marking M' (notation: $M \xrightarrow{t} M'$), defined as follows:*

Let $M(p) = a_1^p a_2^p \ldots a_n^p$ where it holds that $t^-(p) = k$ and $t^+(p) = r$ (For any letter α in an alphabet it holds that $\alpha^0 = \varepsilon$.) Then

$$M'(p) := \begin{cases} a_{k+1}^p \ldots a_n^p 0^r & \text{if } k < n \\ 0^r & \text{if } k = n \end{cases}.$$

Remark 5.8 *Let $M_1 \xrightarrow{t} M_2$ be an admissible state change in $\mathcal{P}$. Then the transition t is enabled in m_{M_1} in $S(\mathcal{P})$ and $m_{M_1} \xrightarrow{t} m_{M_2}$ is an admissible state change in $S(\mathcal{P})$.*

Before introducing state change through elapsing of time we define a modified "modulo"[2] operation.

Definition 5.9 (modified modulo) *Let a and b be natural numbers. We define:*

$$a \widehat{\bmod} b := \begin{cases} a \bmod b & \text{if } a \bmod b \neq 0 \\ b & \text{if } a \bmod b = 0 \end{cases}.$$

---

[2] $a \bmod b := c$ for natural numbers a and $b \neq 0$ with $a = n \cdot b + c$ and $n \in \mathbb{N}, 0 \leq c < b$.

The modified modulo function only differs from the usual modulo definition if a is a multiple of b. In that case the modified modulo function takes the value b instead of zero. So for $b = 0$, where the usual modulo is undefined the new function has the value zero.

Definition 5.10 (elapsing of time) *Let $\mathcal{P}$ be a tw-Petri net and M a time-marking in $\mathcal{P}$. Furthermore, let τ be a non-negative real number. The elapsing of time τ starting in M is always possible (notation: $M \xrightarrow{\tau}$). After time τ has elapsed the time-marking M of $\mathcal{P}$ changes into the time-marking M' (notation: $M \xrightarrow{\tau} M'$), defined as follows:*

Let $M(p) = a_1^p a_2^p \ldots a_n^p$. Let the index i be such that $1 \leq i \leq n$ and

$$u_p < (a_i^p + \tau)\widehat{mod}\, u_p \quad but \quad (a_{i+1} + \tau)\widehat{mod}\, u_p \leq u_p.$$

Then it holds that $M'(p) = b_1^p b_2^p \ldots b_n^p$ with

$$b_j^p := \begin{cases} (a_{i+j}^p + \tau)\,\widehat{mod}\, u_p & if \ i+j \leq n \\ (a_{i+j-n}^p + \tau)\,\widehat{mod}\, u_p & otherwise \end{cases}.$$

Example 5.11 *Let $\mathcal{P}$ be a tw-Petri net and M a time-marking in $\mathcal{P}$ with*

$$M(p) = 3.7\ 2.8\ 2.3\ 2\ 1.5\ 0.3\ 0.1 \ and \ I(p) = (2, 6).$$

Then, for the time-marking M_i' with $M \xrightarrow{\tau_i} M_i'$ the following holds for the place p:

(1) If $\tau_1 = 4$, then

$$M_1'(p) = 6\ 5.5\ 4.3\ 4.1\ 1.7\ 0.8\ 0.3.$$

(2) If $\tau_2 = 14$, then

$$M_2'(p) = 5.7\ 4.8\ 4.3\ 4\ 3.5\ 2.3\ 2.1.$$

(3) If $\tau_3 = 17$, then

$$M_3'(p) = 5.3\ 5.1\ 2.7\ 1.8\ 1.3\ 1\ 0.5.$$

The notions of *firing sequences*, *runs* and *feasible runs* in tw-Petri nets are defined similarly to the respective notions for Time and Timed Petri nets.

5.2   Reachability

In this section we study the reachability behavior of tw-Petri nets and compare it with that of their skeletons.

Theorem 5.12 *Let $\mathcal{P}$ be a tw-Petri net and $S(\mathcal{P}) = (P, T, F, V, m_0)$ its skeleton. Then a transition sequence σ is a firing sequence in $S(\mathcal{P})$ if and only if σ is a firing sequence in $\mathcal{P}$.*

Proof:

($\Longrightarrow$) Let $\sigma = t_1 t_2 \ldots t_n$ be a firing sequence in $S(\mathcal{P})$. Then it holds that

$$m_0 \xrightarrow{t_1} m_1 \xrightarrow{t_2} m_2 \ldots \xrightarrow{t_n} m_n \quad \text{in} \quad S(\mathcal{P}).$$

We will show the existence of a feasible run $\sigma(\tau) = \tau_0 t_1 \tau_1 t_2 \tau_2 \ldots \tau_{n-1} t_n$ in $\mathcal{P}$.

The *idea of the proof* is to use the *ultimo rule*, according to which after each firing we let time elapse until the clock of every token reaches the upper bound of the place the token is lying on. This property is called the *ultimo property* for the related time-marking. At that point each enabled transition is also ready to fire. The ultimo rule is realized by setting the time α between the firing of two transitions to the LCM of all $u_p \in \mathbb{N}^+$ (that is, all u_p that are not ∞ and not zero). This "waiting" time can also be chosen smaller, as we will see in the proof.

The proof is formally done by induction on n:

Basis: $n = 1$, i.e., $\sigma = t_1$.

Let $\tau_0 := \text{LCM} \{u_p \mid p \in P \wedge u_p \neq \infty \wedge u_p \neq 0 \wedge M_0(p) \neq \varepsilon\}$. We consider M_0' with

$$M_0 \xrightarrow{\tau_0} M_0'.$$

For M' it holds that

$$M_0'(p) := \begin{cases} \varepsilon & \text{if } m_0(p) = 0 \\ u_p^{m_0(p)} & \text{otherwise} \end{cases},$$

i.e., M_0' has the ultimo property and it holds that $m_{M_0} = m_0 = m_{M_0'}$. Hence, the transition t_1 is ready to fire in the time-marking M_0'. Let

$$M_0' \xrightarrow{t_1} M_1.$$

Then

$$M_1(p) = \begin{cases} \varepsilon & \text{if } m_1(p) = 0 \\ u_p^{m_0(p) - |t_1^-(p)|} 0^{|t^+(p)|} & \text{if } m_1(p) \neq 0 \end{cases}.$$

Hence, the clock of each token on a place $p \in P$ is zero if the token has arrived there after the last firing or u_p otherwise. We will call this the *pre-ultimo property* for the time-marking and will show that it holds after every firing in the constructed run.

Finally, by Remark 5.8 it still holds that $m_1 = m_{M_1}$.

Step: Let $\sigma = \underbrace{t_1 t_2 \ldots t_{n-1}}_{=:\theta} t_n.$

By *induction hypothesis* it holds that $\theta(\tau) = \tau_0 t_1 \tau_1 t_2 \tau_2 \ldots \tau_{n-2} t_{n-1}$ is already a feasible run in $\mathcal{P}$, so we let

$$M_0 \xrightarrow{\theta(\tau)} M_{n-1}$$

with $m_{n-1} = m_{M_{n-1}}$ such that the pre-ultimo property holds for M_{n-1}. We will show that

$$M_{n-1} \xrightarrow{\tau_{n-1}} M_{n-1}' \xrightarrow{t_n} M_n$$

is feasible in $\mathcal{P}$ and that every token in M_n still has the ultimo property.

Let

$$\tau_{n-1} := \mathrm{LCM} \ \{u_p \mid p \in P \wedge u_p \neq \infty \wedge u_p \neq 0 \wedge M_{n-1}(p) \neq \varepsilon\}.$$

According to Definition 5.10 it holds for every place p and the time-marking M_{n-1}' that $M_{n-1}'(p) \in (u_p)^*$. In other words: M_{n-1}' has the ultimo property. Therefore and by the induction hypothesis it follows that t_n is ready to fire in M_{n-1}'.

Hence,

$$\theta(\tau)\tau_{n-1} t_n$$

is a feasible run of σ in $\mathcal{P}$. Moreover, for the time-marking M_n it holds that $m_n = m_{M_n}$. Then, because of the pre-ultimo property for M_{n-1} and according to Definition 5.7 the time-marking M_n also has the pre-ultimo property.

($\Longrightarrow$) This follows immediately from Remark 5.8. $\qquad\qquad\qquad\qquad$ □

The "waiting time" τ_i can possibly be reduced if there are places which in the time-marking M_i contain tokens but have no post-transitions in the net. We can modify the algorithm from Theorem 5.12 not to take into account the upper bounds of such places in the computation of τ_i. The next example illustrates this slightly modified algorithm.

Example 5.13 *Let us consider the tw-Petri net* $\mathcal{P}_2$ *and the transition sequence* $\sigma = t_1 t_2 t_3$ *in its skeleton* $S(\mathcal{P}_2)$.

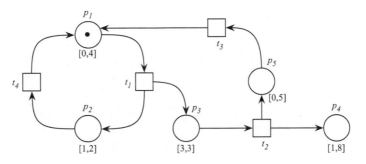

Figure 5.2: The tw-Petri net $\mathcal{P}_2$

We can easily see that σ is a firing sequence in $S(\mathcal{P}_2)$. In order to see that σ is also a firing sequence in $\mathcal{P}$ we must give a feasible run $\sigma(\tau)$ in $\mathcal{P}$, i.e., a $\sigma(\tau)$ with $M_0 \xrightarrow{\sigma(\tau)}$ in $\mathcal{P}$. According to the modified algorithm the run starts with $\tau_0 = 4$, because in M_0 only the place p_1 contains a token and $u_{P_1} = 4$. After four time units have passed, t_1 fires. Now the places p_2 and p_3 are marked and they both have post-transitions. This results in $\tau_1 = LCM\{2,3\} = 6$. After then firing t_2, the places p_2, p_4 and p_5 are marked. For computing τ_2, only p_2 and p_5 are relevant, because p_4 has no post-transitions. Thus, we obtain $\tau_2 = 10$. Therefore we have found the run

$$4\ t_1\ 6\ t_2\ 10\ t_3$$

which is a feasible run of σ in $\mathcal{P}_2$. The algorithm in Theorem 5.12 would have given us the feasible run $\sigma(\tau) = 4\ t_1\ 6\ t_2\ 40\ t_3$.

The following proposition follows immediately from Theorem 5.12.

Proposition 5.14 *Let $\mathcal{P}$ be a tw-Petri net and T its set of transitions. Furthermore, let $\sigma(\tau) = \tau_0 t_1 \tau_1 t_2 \tau_2 \ldots \tau_{n-1} t_n$ with $\tau_i \in \mathbb{R}_0^+$ for $0 \leq i \leq n-1$ be a feasible run in $\mathcal{P}$. Then there also exists a feasible run $\sigma(\tau^*) = \tau_0^* t_1 \tau_1^* t_2 \tau_2^* \ldots \tau_{n-1}^* t_n$ with $\tau_i^* \in \mathbb{N}$ for $0 \leq i \leq n-1$ in $\mathcal{P}$.*

The *state space* $R_{\mathcal{P}}$ of a tw-Petri net $\mathcal{P}$ is the set of all reachable time-markings in the net. It is clear that for an arbitrary marking m^* in the skeleton $S(\mathcal{P})$ the set $\{M \mid M \in R_{\mathcal{P}} \land m_M = m^*\}$ is generally infinite.

Let us consider the set $R_{|\mathcal{P}|} := \{m_M \mid M \in R_{\mathcal{P}}\}$ of all markings m_M in $S(\mathcal{P})$ where M is a reachable time-marking in $\mathcal{P}$. $R_{|\mathcal{P}|}$ corresponds to the set of time-markings in $\mathcal{P}$ when the age of the tokens is ignored.

Definition 5.15 (boundedness) *A tw-Petri net $\mathcal{P}$ is bounded if the set $R_{|\mathcal{P}|}$ is finite.*

The next two assertions follow immediately from Theorem 5.12.

Corollary 5.16 *Let $\mathcal{P}$ be a tw-Petri net and $S(\mathcal{P})$ its skeleton. Then*

(1) $R_{S(\mathcal{P})} = R_{|\mathcal{P}|}$ and

(2) $\mathcal{P}$ is bounded if and only if $S(\mathcal{P})$ is bounded.

Although the sets $R_{S(\mathcal{P})}$ and $R_{|\mathcal{P}|}$ are equal, the reachability behavior of the tw-Petri net can be different from that of its skeleton. The reason for this is that for a firing sequence in the skeleton, say $\sigma = \sigma_1 \sigma_2$, there might be some run of σ_1 in $\mathcal{P}$ which leaves the tw-Petri net in a time-marking in which transitions are enabled but none of them are ready to fire. It can happen that in this last time-marking it is only possible for time to elapse but none of the enabled transitions ever become ready to fire. Note that there is of course always at least one feasible run of σ in $\mathcal{P}$.

Example 5.17 *Let us consider the skeleton of the tw-Petri net* $\mathcal{P}_3$ *given in Fig. 5.3. The transition sequence* $\sigma = \underbrace{t_1\, t_1}_{\sigma_1}\ \underbrace{t_2}_{\sigma_2}$ *is a firing sequence in the Petri net* $S(\mathcal{P}_3)$.

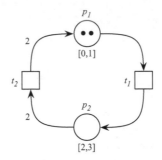

Figure 5.3: The tw-Petri net $\mathcal{P}_3$

The run $\sigma_1(\tau) = 1.5\, t_1\, 1.5\, t_1$ *is a feasible run in* $\mathcal{P}_3$, *but after its firing only time can elapse and* t_2 *can never become ready to fire even though it is always enabled.*
We note that the Petri net $S(\mathcal{P}_3)$ *is live.*

In order to study this kind of a behavior we define the notion of a *time-deadlock*. A transition is in a time-deadlock in a certain time-marking if it is enabled but not ready to fire in the time-marking and no elapse of time can change this. In other words: The time restrictions prevent the firing of the transitions.

Definition 5.18 (time-deadlock) *Let* $\mathcal{P}$ *be a tw-Petri net and* M *a time-marking in* $\mathcal{P}$. *The transition* $\hat{t}$ *is in a time-deadlock in* M *if it holds that:*

1. $\hat{t}^- \leq m_M$ *and*

2. $\forall \tau (\tau \in \mathbb{R}_0^+ \longrightarrow M \xrightarrow{\tau} \overset{\hat{t}}{\nrightarrow})$.

If a tw-Petri net is in a time-marking in which all enabled transitions are in time-deadlocks it is obviously only possible for time to pass. This means in any extension of this run no more transitions can fire, even though the respective transition sequence can be continued in the skeleton.

The following Example 5.19 shows that the usual discretization of the state space based on all possible integer time elapses is generally not sufficient for studying the behavior of a tw-Petri net. To see this we define the *integer reachability graph* $\mathcal{IR}_\mathcal{P}$ of a tw-Petri net $\mathcal{P}$, a labeled digraph whose vertices are the reachable integer time-markings in $\mathcal{P}$ and whose edges are triples (M, t, M') or $(M, 1, M')$ such that $M \xrightarrow{t} M'$ or $M \xrightarrow{1} M'$ is feasible in $\mathcal{P}$, respectively. An edge (M, t, M') is labeled with t and an edge $(M, 1, M')$ with 1.

Example 5.19

Let us again consider the tw-Petri net $\mathcal{P}_3$ shown in Fig. 5.3.

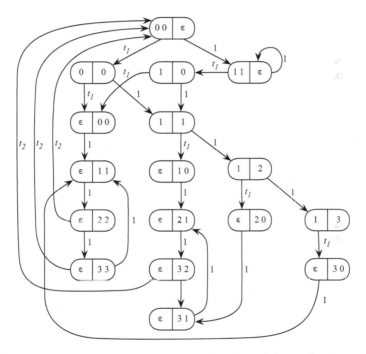

Figure 5.4: The integer reachability graph $\mathcal{IR}_{\mathcal{P}_3}$ of the tw-Petri net $\mathcal{P}_3$

As we have already seen, after firing the run $\sigma_1(\tau) = 1.5\ t_1\ 1.5\ t_1$ the only enabled transition t_2 is in a time-deadlock. Hence, after reaching this time-marking no firing is possible any more. The integer reachability graph $\mathcal{IR}_{\mathcal{P}_3}$ on the other hand is strongly connected (cf. Fig. 5.4) and therefore every feasible run of $\mathcal{P}$ represented here can be continued infinitely.

A further characteristic property of tw-Petri nets are possible "time-gaps" in firing sequences. This means that the time continuum $[0, \infty]$ can be divided into intervals $[0, a_0], [a_0, a_1], [a_1, a_2], [a_2, a_3], [a_3, a_4], \ldots$ such that it is possible to find a feasible run $\sigma(\tau^i)$ of the transition sequence σ with a length $\ell(\sigma(\tau^i)) \in [a_i, a_{i+1}]$ for any $i = 0, 2, 4, \ldots$ but there is no feasible run with time-length from $[0, a_0]$ or from $[a_i, a_{i+1}]$ for any $i = 1, 3, 5, \ldots$.

Such time-gaps are impossible in Time Petri nets, but there are tw-Petri nets with time-gaps. The next proposition proves the absence of time-gaps in Time Petri nets. We will then give an example of a tw-Petri net with time-gaps.

Proposition 5.20 *Let $\mathcal{Z}$ be a TPN and $\sigma = t_1 t_2 \ldots t_n$ an arbitrary transition sequence in $\mathcal{Z}$. Furthermore, let $\sigma(\tau_\alpha) = \tau_0^\alpha\ t_1\ \tau_1^\alpha\ t_2\ \tau_2^\alpha \ldots \tau_{n-1}^\alpha\ t_n\ \tau_n^\alpha$ and $\sigma(\tau_\beta) = \tau_0^\beta\ t_1\ \tau_1^\beta\ t_2\ \tau_2^\beta \ldots \tau_{n-1}^\beta\ t_n\ \tau_n^\beta$ be two feasible runs of the transition sequence σ, with $\ell(\sigma(\tau_\alpha)) = \alpha$ and $\ell(\sigma(\tau_\beta)) = \beta$ and $\alpha < \beta$. Then for each $\gamma \in [\alpha, \beta]$ there exists a feasible run $\sigma(\tau_\gamma) = \tau_0^\gamma\ t_1\ \tau_1^\gamma\ t_2\ \tau_2^\gamma \ldots \tau_{n-1}^\gamma\ t_n\ \tau_n^\gamma$ with $\ell(\sigma(\tau_\gamma)) = \gamma$.*

Idea of the proof. Let us consider the parametric run $(\sigma(x), B_\sigma)$.

Then, the length of every run $\sigma(\tau)$ of the transition sequence σ is given by the linear function

$$f(x) := \sum_{i=0}^{n} x_i \quad \text{such that } x \in \mathbb{R}^{n+1} \text{ and } x \text{ satisfies } B_\sigma.$$

Let $H_\sigma := \{x \in \mathbb{R}^{n+1} \mid x \text{ satisfies } B_\sigma\}$. The length of $\sigma(\tau_\alpha)$ is then $f(x) = \alpha$ for $x = \tau_\alpha$ and the length of $\sigma(\tau_\beta)$ is $f(x) = \beta$ for $x = \tau_\beta$. Obviously the values α and β are real numbers between $\mu = \min\{f(x) \mid x \in H_\sigma\}$ and $\lambda = \max\{f(x) \mid x \in H_\sigma\}^3$.

---

[3]If λ exists. Otherwise α and β are not bounded from above.

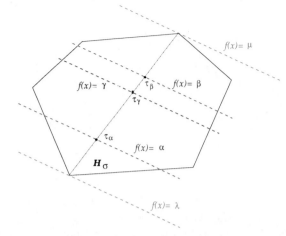

Figure 5.5: Graphical illustration of H_σ for $n = 2$

Because the polyhedron H_σ is convex and the points τ_α and τ_β belong to H_σ, there is a point τ_γ, which belongs to H_σ (to the segment defined by the points τ_α and τ_β), and lies on some hyperplane parallel to the hyperplanes $f(x) = \alpha$ and $f(x) = \beta$ (cf. [GLS93]). This means that $f(x) = \gamma$ for $x = \tau_\gamma$. □

In general, not all tw-Petri nets have time-gaps, but it is possible to find a tw-Petri net $\mathcal{P}$ and a firing sequence σ with two feasible runs in $\mathcal{P}$ whose lengths are α and β such that no run of σ with length γ, $\alpha < \gamma < \beta$ exists. This of course is neither true for all α and β nor for all firing sequences.

The next example verifies this fact.

Example 5.21 *Let us consider the tw-PN $\mathcal{P}_4$ given in Fig. 5.6 and the transition sequence $\sigma_1 = t_1 t_2 t_3$. The runs $3\,t_1 3\,t_2 3\,t_3$ and $5\,t_1 2\,t_2 3\,t_3$ are feasible runs of σ_1 with lengths 9 and 10. It is easy to see that there does not exist a run of σ_1 whose length is, e.g., 9.5 or any other number strictly between 9 and 10. The lengths of all feasible runs of σ_1 belong to the intervals $[7, 9], [10, 12], [13, 15], \ldots$.*

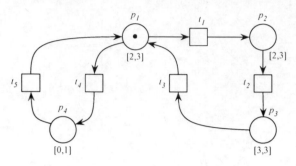

Figure 5.6: The tw-Petri net $\mathcal{P}_4$

Considering the sequence $\sigma_2 = t_4 t_5 t_4$, it is obvious that the shortest time length of a feasible run of σ_2 is 4. Furthermore, it can easily be seen that for each real number $k \geq 4$ a feasible run of σ_2 with time length k exists.

5.3   Petri Nets with Retention Time Places and Counter Machines

In Chapter 2 we showed that classic Petri nets are not Turing equivalent and therefore it is generally not possible to simulate counter machines with such nets. The reason is that classic Petri nets cannot simulate the zero-test, so there is no way to check whether some place in a classic Petri net does not contain any tokens. The explicit addition of time made Time and Timed Petri nets more powerful. The real cause of this gain in power however is not time itself, but the possibility to force transitions to fire at some point in time, either at the end of an interval (Time Petri nets) or immediately after their enabling (Timed Petri nets). The *absence of any compulsion to fire* in tw-Petri nets is the reason why they, too, cannot simulate the zero-test. Let us examine this claim in more detail:

We assume that the zero-test can be simulated with tw-Petri nets, so we can check for a place p whether it is marked or not. There are thus two further places p_{marked} and p_{empty}, both of them empty in the initial marking and if p is marked in M then p_{marked} will be marked after a finite amount of time and p_{empty} will always stay unmarked and if p is empty in M then p_{empty} will be marked after a finite amount of time and p_{marked} will always stay empty.

Case 1: $M_0(p) = \varepsilon$.

In the tw-Petri net $\mathcal{P}$ checking whether p is empty or not it then holds for any feasible run $\sigma(\tau)$ in $\mathcal{P}$ with

$$M_0 \xrightarrow{\sigma(\tau)} M \text{ that}$$
$$M(p_{marked}) = \varepsilon \text{ and } M(p_{empty}) \neq \varepsilon.$$

So let us consider an arbitrary run $\sigma(\tau)$. By Theorem 5.12 it follows that the transition sequence σ is also a firing sequence in the skeleton $S(\mathcal{P})$, i.e.,

$$m_o = m_{M_0} \xrightarrow{\sigma(\tau)} m_M \quad \text{in } S(\mathcal{P}) \text{ and}$$
$$m_{M(p_{marked})} = 0 \text{ and } m_{M(p_{empty})} \neq 0.$$

There are two reasons for p_{marked} to stay empty in $\mathcal{P}$.

It can be impossible to extend the run $\sigma(\tau)$ because of time-deadlocks for all enabled transitions in the time-marking M. If p_{marked} is empty because of time-deadlocks, then we consider the feasible run of σ using the ultimo-rule. During the firing of this run no enabled transition is ever in a time-deadlock. Therefore after firing this feasible run both places p_{marked} and p_{empty} would be marked. This is a contradiction to the assumption that $\mathcal{P}$ can simulate the zero-test for p.

If however there is no feasible run in $\mathcal{P}$ without any time-deadlocks leading to a time-marking such that p_{marked} is marked then, because of Theorem 5.12, there is no firing sequence in $S(\mathcal{P})$ such that after its firing the place p_{marked} is marked and therefore the classic Petri net $S(\mathcal{P})$ can check whether p is empty.

Case 2: $M_0(p) \neq \varepsilon$.

We can similarly conclude that $S(\mathcal{P})$ is able to check whether p is marked.

Thus our assumption lets us prove that the classic Petri net $S(\mathcal{P})$ can simulate the zero-test which contradicts Theorem 2.18. Hence, the zero-test can not be simulated by a tw-Petri net and we have shown the following:

Theorem 5.22 *tw-Petri nets are not Turing equivalent.*

The computational power of tw-Petri nets is however no smaller than that of classic Petri nets, because every classic Petri net can be simulated by a tw-Petri net with the interval $[0, \infty]$ associated with each place.

5.4 Liveness

Liveness in tw-Petri nets can be introduced similarly as for Time Petri nets. We know that every firing sequence of a tw-Petri net is also a firing sequence in its skeleton and vice versa. This results in the same boundedness behavior for a tw-Petri net and its skeleton. Surprisingly however, the liveness behavior of a tw-Petri net and that of its skeleton are not equivalent. This is caused by the existence of time-deadlocks in some tw-Petri nets, meaning the existence of reachable time-markings such that all enabled transitions are in time-deadlocks.

Thus, we can assert the following:

Proposition 5.23 *The skeleton of any live tw-Petri net is live, too.*

Proof: The statement follows immediately from Theorem 5.12.   □

There are however some structurally restricted classes of tw-Petri nets with the same liveness behavior as their skeletons. For more on this cf. [WPZ09].

5.5 Bibliographical Notes

Petri nets in which time is explicitly associated with the places are used less often than Time or Timed Petri nets. This probably results from the fact that the systems that are developed using these nets can also easily be modeled by other time-dependent nets, which are already well studied and for which tools facilitating the analysis are available.

In [JR83] a time-dependent Petri net is introduced in which for each place a time delay is specified. The transitions in the net fire immediately and according to the maximal-step rule. Only 1-safe nets are considered. A general methodology for analyzing them is not presented. These nets were introduced in order to design control systems and to compute time distances between system states. For a subset of this class of time-dependent Petri nets, the marked graphs, a state equation is given using minmax-algebra in [Wan98].

Petri nets with explicit time specifications for places as presented here were first introduced in [PZ93]. In [Lem95] two variants of these time-dependent nets were introduced and thoroughly studied. In the first variant each place has a minimum retention time and there is no compulsion to fire. In the second variant a retention time and a time period of validity are associated with each place and combined with a compulsion to fire. Reachability graphs are used to analyze the qualitative and quantitative properties of such nets.

5.6   Exercises

Exercise 5.1

(1) Are all feasible T-invariants of the skeleton of an arbitrary tw-Petri net
feasible T-invariants of the tw-Petri net? Justify your answer.

(2) Are all P-invariants of the skeleton of an arbitrary tw-Petri net also
P-invariants of the tw-Petri net? Justify your answer.

Exercise 5.2

Are time-gaps possible in Timed Petri nets? If so, give an example and if
not, a proof.

Exercise 5.3

Is there a (structurally) restricted subset of the set of all tw-Petri nets such
that the liveness behavior of such nets is fully represented by their integer
reachability graph even though the time elapses are modeled with ratio-
nal/real numbers? Justify your answer.

Appendix A

A.1 Appendix on Time Petri Nets

The Time Petri nets $\mathcal{Z}_9$ and $\mathcal{Z}_{10}$, as well as the results shown in Tables A.1, A.2, A.3, A.4 and A.5 were first represented in [Pil09]. In the next six tables, the sizes of the reachability graphs, which is defined as the sum of the number of vertices and the number of edges, are compared.

The reachability graphs, given below, were computed with the tools TINA (cf. [Ber09]) and partly with INA (cf. [Sta97]) and Charlie (cf. [Hei11]).

For every Time Petri net, TINA computes the reachability graph based on state classes and the reachability graph based on essential states. The computations are done according to the definition of dynamic conflict (cf. Definition 2.19). Nevertheless, we can use TINA for computations of reachability graphs where the firing rule is defined based on the definition of static conflict. For this reason a place q_i can be added to each shared place p_i in a Time Petri net. The place q_i is marked with a token in the initial marking and q_i is a pre-place and post-place for each post-transition of p_i. The multiplicity of every input-arc and output-arc of q_i is 1. When the firing rule is defined based on static conflict this modification of the net does not change the number of vertices or the number of edges in either reachability graph, but it ensures that every conflict in the net is a static conflict.

INA computes reachability graphs based on essential states where the firing rule is defined based on the definition of static conflict. INA also computes minimum and maximum distances.

Charlie computes reachability graphs based on essential states, according to the definition of either dynamic or static conflict.

The next five tables compare for different initial place-markings the sizes of the two reachability graphs of $\mathcal{Z}_9$, one based on essential states and the other based on state classes.

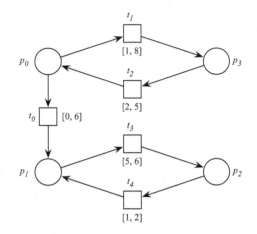

Figure A.1: The Time Petri net $\mathcal{Z}_9$

number of	essential-states algorithm			state class algorithm		
tokens in p_0	number of vertices	number of edges	total number	number of vertices	number of edges	total number
0	1	0	1	1	0	1
1	4	21	25	4	5	9
2	63	310	373	81	157	238
3	250	1252	1502	258	574	832
4	692	3920	4612	1053	2979	4032
5	1367	8115	9482	2653	8119	10772
6	2265	13769	16034	5000	15884	20884
7	3386	20882	24268	8089	26315	34404
8	4730	29454	34184	11909	39371	51280
9	6297	39485	45782	16454	55023	71477
10	8087	50975	59062	21708	73210	94918

Table A.1: A comparison of the size of the reachability graphs for the Time Petri net $\mathcal{Z}_9$ by increasing the number of tokens in place p_0 from 0 to 10. The firing rule is defined based on static conflict.

number of	essential-states algorithm			state class algorithm		
tokens in p_0	number of vertices	number of edges	total number	number of vertices	number of edges	total number
0	1	0	1	1	0	1
1	4	21	25	4	5	9
2	86	441	527	94	186	280
3	550	2740	3290	570	1354	1924
4	1916	9975	11891	2181	5907	8088
5	9167	50618	59785	16588	53781	70369
7	15152	84449	99601	34118	114249	148367
8	22862	127989	150851	61123	208195	269318
9	32165	180510	212675	97479	335218	432697
10	42989	241713	284702	142712	493602	636314

Table A.2: A comparison of the size of the reachability graphs for the Time Petri net $\mathcal{Z}_9$ by increasing the number of tokens in place p_0 from 0 to 10. The firing rule is defined based on dynamic conflict.

number of tokens in p_1	essential-states algorithm			state class algorithm		
	number of vertices	number of edges	total number	number of vertices	number of edges	total number
0	1	0	1	1	0	1
1	2	4	6	2	2	4
2	4	8	12	3	3	6
3	4	8	12	3	3	6
4	4	8	12	3	3	6
5	4	8	12	3	3	6
6	4	8	12	3	3	6
7	4	8	12	3	3	6
8	4	8	12	3	3	6
9	4	8	12	3	3	6
10	4	8	12	3	3	6

Table A.3: A comparison of the size of the reachability graphs for the Time Petri net $\mathcal{Z}_9$ by increasing the number of tokens in place p_1 from 0 to 10. These values are independent of whether the firing rule is defined based on static or dynamic conflict.

number of	essential-states algorithm			state class algorithm		
tokens	number of	number of	total	number of	number of	total
in p_2	vertices	edges	number	vertices	edges	number
0	1	0	1	1	0	1
1	2	4	6	2	2	4
2	4	8	12	3	3	6
3	9	18	27	6	6	12
4	17	35	52	10	11	21
5	26	55	81	16	19	35
6	35	75	110	28	34	62
7	44	95	139	39	51	90
8	53	115	168	53	70	123
9	62	135	197	68	92	160
10	71	155	226	83	114	197
11	80	175	255	98	136	234
12	89	195	284	113	158	271
13	98	215	313	128	180	308
14	107	235	342	143	202	345
15	116	255	371	158	224	382

Table A.4: A comparison of the size of the reachability graphs for the Time Petri net $\mathcal{Z}_9$ by increasing the number of tokens in place p_2 from 0 to 15. These values are independent of whether the firing rule is defined based on static or dynamic conflict.

number of tokens in p_3	essential-states algorithm			state class algorithm		
	number of vertices	number of edges	total number	number of vertices	number of edges	total number
0	1	0	1	1	0	1
1	4	21	25	4	5	9
2	79	383	461	87	169	256
3	523	2521	3044	539	1273	1812
4	1810	9258	11068	2051	5515	7566
5	4182	22274	26456	6051	18277	24328
6	7344	39718	47062	13742	44134	57876
7	11218	61345	72563	25174	83106	108280
8	15996	88100	104096	41145	138054	179199
9	21585	119380	140965	61281	207597	268878
10	28061	155778	183839	86457	294929	381386

Table A.5: A comparison of the size of the reachability graphs for the Time Petri net $\mathcal{Z}_9$ by increasing the number of tokens in place p_3 from 0 to 10. The firing rule is defined based on dynamic conflict.

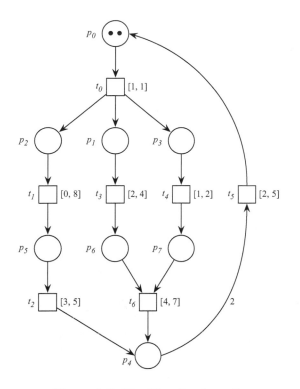

Figure A.2: The Time Petri net $\mathcal{Z}_{10}$

The next table compares the size of the two reachability graphs of $\mathcal{Z}_{10}$ based on essential-states and based on state classes for different initial place-markings and different time intervals for the transition t_0.

number of tokens in p_0	interval $[eft(t_0), lft(t_0)]$	essential-states algorithm			state class algorithm		
		number of vertices	number of edges	total number	number of vertices	number of edges	total number
2	[1,1]	1916	7996	9912	2335	5938	8273
3	[1,1]	32157	147941	180098	97963	321294	419257
3	[1,2]	38708	184806	223514	120115	404507	524622
3	[1,3]	41436	202257	243693	127228	433980	561208
3	[1,4]	43114	213734	256848	125887	431067	556954
3	[1,5]	44181	222022	266203	122693	418867	541560
3	[1,6]	44923	228699	273622	119489	406725	526212

Table A.6: A comparison of the size of the reachability graphs for the Time Petri net Z_{10} by increasing the number of tokens in place p_0 from 2 to 3 and the $lft(t)_0$ from 1 to 6. These values are independent of whether the firing rule is defined based on static or dynamic conflict.

Bibliography

[AD94]   Rajeev Alur and David L. Dill. A theory of timed automata. *Theoretical Computer Science*, 126(2):183–235, 1994.

[Bac11]   J.P. Bachmann. Zeitunabhängige Lebendigkeit von Intervall–Petrinetzen (Diplomarbeit, in German). Master's thesis, Humboldt-Universität zu Berlin, Germany, 2011.

[BB93]   H. Boucheneb and G. Berthelot. Towards a Simplified Building of Time Petri Net Reachability Graphs. In *Proceedings of Petri Nets and Performance Models PNPM 93*, Toulouse, France, 1993. IEEE Computer Society Press.

[BB94]   G. Berthelot and H. Boucheneb. Occurrence Graph for Interval Timed Colored Nets. In *Application and Theory of Petri Nets*, LNCS, pages 79–98. Springer, 1994.

[BCH+13]   B. Bérard, F. Cassez, S. Haddad, D. Lime, and O. H. Roux. The expressive power of time Petri nets. *Theor. Comput. Sci.*, 474:1–20, February 2013.

[BD91]   B. Berthomieu and M. Diaz. Modeling and Verification of Time Dependent Systems Using Time Petri Nets. In *Advances in Petri Nets 1984*, volume 17, No. 3 of *IEEE Trans. on Software Eng.*, pages 259–273, 1991.

[Ber09]   B. Berthomieu. *TIme petri Net Analyzer*. LAAS / CNRS, Toulouse, France, http://www.laas.fr/bernard/tina/, 2.9.8 released edition, 2009.

[Bes87]   E. Best. Structure Theory of Petri Nets: the Free Choice Hiatus. *LNCS*, 254:168–205, 1987.

[BFSV04]   G. Bucci, A. Fedeli, L. Sassoli, and E. Vicario. Timed State Space Analysis of Real-Time Preemptive Systems. *IEEE Transactions on Software Engineering*, 30(2):97–111, 2004.

[BK02]   F. Bause and P. S. Kritzinger. *Stochastic Petri Nets - An Introduction to the Theory*. Friedr. Vieweg & Sohn Verlag, second edition, 2002.

[BM83]   B. Berthomieu and M. Menasche. An Enumerative Approach for Analyzing Time Petri Nets. In *Proceedings IFIP Congress*, September 1983.

[BPZ10]   J.P. Bachmann and L. Popova-Zeugmann. Time-Independent Liveness in Time Petri Nets. *Fundamenta Informaticae (FI),102,IOS Press, Amsterdam*, pages 1–17, 2010.

[CHEP71]   F. Commoner, A. Holt, S. Even, and A. Pnueli. Marked directed graphs. In *Journal of Computer and System Science*, volume 5, pages 511–523, 1971.

[CLRS01]   Thomas H. Cormen, Charles E. Leiserson, Ronald L. Rivest, and Clifford Stein. *Introduction to Algorithms*. MIT Press, second edition, 2001.

[Com73]   F.G. Commoner. Deadlocks in Petri Nets. Report CA-7206-2311, Applied Data Research, New York, 1973.

[CR04]   Franck Cassez and Olivier H. Roux. Structural translation from time Petri nets to timed automata. In *Fourth International Workshop on Automated Verification of Critical Systems (AVoCS'04)*, Electronic Notes in Theoretical Computer Science, London (UK), September 2004. Elsevier.

[Esp98]   J. Esparza. Decidability and Complexity of Petri Net Problems – an Introduction. In G. Rozenberg and W. Reisig, editors, *Lectures on Petri Nets I: Basic Models. Advances in Petri Nets*, number 1491 in Lecture Notes in Computer Science, pages 374–428, 1998.

[Fin93]   A. Finkel. The Minimal Coverability Graph for Petri Nets. In Rozenberg, editor, *Lecture Notes in Computer Science*, volume 674, pages 210–243. Springer-Verlag, 1993.

[Gen68]   H. J. Genrich. Das Zollstationenproblem. Interner Bericht GMD/I5, GMD, Bonn, 1968.

[GLS93]   Martin Grötschel, Lásló Lovász, and Alexander Schrijver. *Geometric Algorithms and Combinatorial Optimization*. Springer-Verlag, second corrected edition, 1993.

[Hac72]   M. H. T. Hack. Analysis of Production Schemata by Petri-Nets. Project MAC, MIT, 1972.

[Hei11]   M. Heiner. *Charlie*. Brandenburgische Technische Universität, http://www-dssz.informatik.tu-cottbus.de/DSSZ/Software/Charlie, January 2011.

[HMU02]   John E. Hopcroft, R. Motwani, and J. D. Ullman. *Einführung in die Automatentheorie, formale Sprachen und Komplexitätstheorie*. Pearson Studium, 2002.

[HPZ97]   M. Heiner and L. Popova-Zeugmann. Worst-case Analysis of Concurrent Systems with Duration Interval Petri Nets. In *Tagungsband zur 5. Fachtagung Entwurf komplexer Automatisierungssysteme*, 1997.

[JR83]   James E. Coolahan Jr. and Nick Roussopoulos. Timing Requirements for Time-Driven Systems Using Augmented Petri Nets. *IEEE Trans. Software Eng.*, 9(5):603–616, 1983.

[KM69]   R. M. Karp and R. E. Miller. Parallel program schemata. *Journal of Computer and System Science*, 3:147–195, 1969.

[Lau73]   Kurt Lautenbach. *Exakte Bedingungen der Lebendigkeit für eine Klasse von Petri-Netzen*. Dissertation, GMG, Bonn, 1973.

[Lem95]   Karsten Lemmer. *Diagnose diskreter modellierter Systeme mit Petrinetzen*. Dissertation, Technische Universität Braunschweig, 1995.

[May80]   E. W. Mayr. Ein Algorithmus für das allgemeine Erreichbarkeit-
          sproblem bei Petrinetzen und damit zusammenhängende Prob-
          leme. Technischer Bericht der TU München TYM-I8010 (Disser-
          tation), Institut für Informatik, 1980.

[MBC+96] M. Ajmone Marsan, G. Balbo, G. Conte, S. Donatelli, and
          G. Franceschinis. *Modelling with Generalized Stochastic Petri
          Nets*. John Wiley and Sons, 1996.

[Mer74]   Philip M. Merlin. *A Study of the Recoverability of Computing
          Systems*. PhD thesis, University of California, Irvine, January
          1974.

[MF76]   Philip M. Merlin and David J. Farber. Recoverability of Commu-
          nication Protocols–Implications of a Theoretical Study. *IEEE
          Transactions on Communications*, vol. 24, no. 9:1036 – 1043,
          September 1976.

[Mur89]   Tadao Murata. Petri Nets: Properties, Analysis and Applica-
          tions. In *Proceedings of the IEEE*, volume 77(4), pages 541–580,
          1989.

[Pen00]   W. Penczek. Partial Order Reductions for Checking Branching
          Properties of Time Petri Nets . *Proc. of the Int. Workshop on
          CS&P'00 Workshop, Informatik-Berichte No.140(2)*, pages 189–
          202, 2000.

[Pet62]   C. A. Petri. Kommunikation mit Automaten. Schriften des
          Instituts für Instrumentelle Mathematik No. 2, Bonn, 1962.

[Pet77]   James L. Peterson. Petri nets. *ACM Comput. Surv.*, 9(3):223–
          252, 1977.

[Pet81]   James Lyle Peterson. *Petri Net Theory and the Modeling of Sys-
          tems*. Prentice Hall PTR, Upper Saddle River, NJ, USA, 1981.

[Pil09]   A. Pilchowski. Vergleich zweier Algorithmen zur Erzeugung von
          Erreichbarkeitsgraphen von Intervall-Petrinetzen hinsichtlich der
          Größe ihrer resultierenden Graphen (Diplomarbeit, in German).
          Master's thesis, Humboldt-Universität zu Berlin, Germany, 2009.

[Pop91]   L. Popova. On Time Petri Nets. *J. Inform. Process. Cybern. EIK* *27(1991)4*, pages 227–244, 1991.

[PP06]   Wojciech Penczek and Agata Polrola. *Advances in Verification of* *Time Petri Nets and Timed Automata.* Springer-Verlag, Berlin, 2006.

[PS98]   Ch. Papadimitriou and K. Steiglitz. *Combinatorial Optimization:* *Algorithms and Complexity.* Dover Publications, Inc., Mineola, New York, 1998.

[PW03]   L. Priese and H. Wimmel. *Theoretische Informatik Petri-Netze.* Springer-Verlag, Berlin Heidelberg New York, 2003.

[PZ89]   L. Popova-Zeugmann. *Zeit-Petri-Netze.* Dissertation, Humboldt-Universität zu Berlin, 1989.

[PZ93]   L. Popova-Zeugmann. Petri Nets with Time Restrictions. *Systems* *Analysis, Modeling, Simulation (SAMS)*, pages 13–20, 1993.

[PZ07]   L. Popova-Zeugmann. Time Petri Nets State Space Reduction Using Dynamic Programming. *Journal of Control and Cybernetics*, 35(3):721–748, 2007.

[PZ11]   L. Popova-Zeugmann. Quantitative evaluation of time-dependent Petri nets and applications to biochemical networks. *Natural* *Computing*, 10(3):1017–1043, 2011.

[PZHK05]   L. Popova-Zeugmann, M. Heiner, and I. Koch. Time Petri Nets for Modelling and Analysis of Biochemical Networks. *Funda-* *menta Informaticae (FI) 67, IOS Press, Amsterdam*, pages 149–162, 2005.

[PZP12]   L. Popova-Zeugmann and E. Pelz. Algebraical Characterisation of Interval-Timed Petri Nets with discrete delays. *Fundamenta In-* *formaticae (FI), Volume 120, IOS Press, Amsterdam*, pages 341–357, 2012.

[PZS99]   L. Popova-Zeugmann and D. Schlatter. Analyzing Path in Time Petri Nets. *Fundamenta Informaticae (FI) 37, IOS Press, Ams-* *terdam*, pages 311–327, 1999.

[Ram74]   C. Ramchandani. Analysis of Asynchronous Concurrent Systems by Timed Petri Nets. Technical Report: TR-120, Massachusetts Institute of Technology, Cambridge, MA, USA, February 1974.

[Rei13]   W. Reisig. *Understanding Petri Nets: Modeling Techniques, Analysis Methods, Case Studies*. Springer Verlag, April 2013. 230 pages; ISBN 978-3-642-33277-7.

[Ric06]   Jan Richling. *Komponierbarkeit eingebetteter Echtzeitsysteme*. Dissertation, Humboldt-Universität zu Berlin, 2006.

[RPZW02] J. Richling, L. Popova-Zeugmann, and M. Werner. Verification of Non-functional Properties of a Composable Architecture with Petrinets. *Fundamenta Informaticae (FI) 51, IOS Press, Amsterdam*, pages 185–200, 2002.

[RWPZ02] J. Richling, M. Werner, and L. Popova-Zeugmann. Automatic composition of timed petrinet specifications for a real-time architecture. In *Proceedings of 2002 IEEE International Conference on Robotics and Automation*, Washington, D.C., May 2002.

[Sif77]   J. Sifakis. Use of Petri nets for Performance Evaluation. *3rd Intl. Symposium on Modeling and Evaluation, IFIP, North-Holland*, pages 75–93, 1977.

[Sif80]   J. Sifakis. Performance evaluation of systems using nets. In Brauer, W., editor, *Lecture Notes in Computer Science: Net Theory and Applications, Proc. of the Advanced Course on General Net Theory of Processes and Systems, Hamburg, 1979*, volume 84, pages 307–319, Berlin, Heidelberg, New York, 1980. Springer-Verlag.

[Sta80]   P.-H. Starke. *Petri-Netze*. Deutscher Verlag der Wissenschaften, Berlin, 1980.

[Sta87]   Peter H. Starke. Remarks on Timed Nets. *Petri Net Newsletter*, 27:37–47, August 1987.

[Sta90]   Peter H. Starke. *Analyse von Petri-Netz-Modellen*. B.G. Teubner, Stuttgart, 1990.

[Sta95]   Peter H. Starke. A Memo on Time Constraints in Petri Nets. Informatik-Berichte, No. 46, Humboldt Universität zu Berlin, Inst. für Informatik, August 1995.

[Sta97]   P. H. Starke. *INA - Integrated Net Analyzer*. Berlin, 1997. Manual.

[Wal82]   B. Walter. *Transaktionsorientierte Recovery-Konzepte für verteilte Datenbanksysteme*. Dissertation, Universität Stuttgart, 1982.

[Wan98]   Jiacun Wang. *Timed Petri Nets*. Kluwer Academic Publishers, Boston, Dordrecht, London, 1998.

[WPZ09]   J.-Th. Wegener and L. Popova-Zeugmann. Petri Nets with Time Windows: A Comparison to Classical Petri Nets. *Fundamenta Informaticae (FI), 39(2009), IOS Press, Amsterdam*, pages 337–352, 2009.

[WPZR04] M. Werner, L. Popova-Zeugmann, and J. Richling. A Method to Prove Non-Reachability in Priority Duration Petri Nets. *Fundamenta Informaticae (FI),61,IOS Press, Amsterdam*, 2004.

Index

$REIS_{\mathcal{Z}}$, $\underline{85}$, 86
$RIS_{\mathcal{Z}}$, $\underline{52}$, 71, 73
$RS_{\mathcal{Z}}$, 38
$R_{|\mathcal{P}|}$, 181
$R_{\mathcal{N}}$, 13
$R_{\mathcal{Z}}$, 38
$S(\mathcal{D})$, 140
$S(\mathcal{P})$, 174
$S(\mathcal{Z})$, 32
$\mathcal{CG}_{\mathcal{N}}$, 15
$\mathcal{IR}_{\mathcal{P}}$, 183
$\mathcal{M}ax(p)$, 98
$\mathcal{M}in(p)$, 98
$\mathcal{RG}_{\mathcal{D}}$, 145
$\mathcal{RG}_{\mathcal{N}}$, 14
$\mathcal{RG}_{\mathcal{Z}}$, 72, 77
$\mathcal{RG}_{\mathcal{Z}}^{w}$, 129
$\mathcal{RG}_{\mathcal{Z}}^{redu}$, 75, 87

AC net, $\underline{21}$, 112

bag matrix, 156
BFC net, $\underline{112}$, 113
blocking-free, $\underline{22}$, 23, 94
boundedness
    in a Petri net, 13
    in a Time Petri net, 38
    in a Timed Petri net, 145
    in a tw-Petri net, 181

Computability, 17

computable
    CM-, 40
    PN-, 18
    Timed PN-, 148
    TPN-, 42
    Turing-, 17
conflict
    dynamic, $\underline{21}$, 191
    static, $\underline{21}$, 191

distance of time
    between p-markings, $\underline{121}$, 130, 132
    between states, $\underline{119}$, 129, 131

EFC net, $\underline{21}$, 99
elapsing of time
    in a Time Petri net, 36
    in a Timed Petri net, 144
    in a tw-Petri net, 177
enabled, 11

FC net, 21
firing
    in a Petri net, 11
    in a Time Petri net, 35
        modified, 78
    in a Timed Petri net, 144
    in a tw-Petri net, 176
firing sequence
    in a Petri net, 11
    in a Time Petri net, 38

in a Timed Petri net, 155
in a tw-Petri net, 177
function
   duration, 140, 149
   interval, 32, 33, 124, 126, 127
   number-theoretical, 6, 39, 42, 148
   priority, 166

global reachability, 155
global step, 154

homogeneous, 21, 99, 113

incidence matrix, 10, 24
   time-extended, 156

liveness
   in a Petri net, 22, 23
   in a Time Petri net, 94,
      96–98, 113
   in a Timed Petri net, 145
   in a tw-Petri net, 188

marked graph, 21
marking, 8
   t^+, 10
   t^-, 10
   generalized, 14
   initial, 8, 9
   reachable, 10, 13, 13
maximal step, 4, 141, 142
   prioritized, 166
modified modulo, 176
multiset, 154

P-invariant, 25
Parikh matrix, 157
Parikh vector, 12, 24, 118
Petri net, 7

Petri net with retention time on places,
   see tw-Petri net
place-marking (p-marking), 33, 34, 40,
   45, 92
   reachable, 38, 51, 67, 71, 78, 85,
      86, 110
progress matrix, 157

reachability graph
   of a Petri net, 14
   of a Time Petri net, 72, 75, 87,
      191
   of a Timed Petri net, 145
   of a tw-Petri net, 183
ready to fire
   in a Time Petri net, 35
   in a tw-Petri net, 176
run, 37
   feasible, 37
   integer-, 52, 67
   length of a, 118
   maximum, 119, 121
   minimum, 119, 121
   parametric, 46

skeleton
   of a Time Petri net, 32
   of a Timed Petri net, 140
   of a tw-Petri net, 174
state
   in a Time Petri net, 34
     essential-, 85
     initial, 34, 124
     integer-, 51, 67
     parametric, 46
     reachable, 38
   in a Timed Petri net, 141
state class, 49

state equation
   in a Petri net, 12, 151
   in a Timed Petri net, 158,
      162

T-invariant, 24, 118
time dimension, 152
Time Petri net, 32
   finite, 71
   infinite, 78
   lazy, 92, 93, 95, 97
   speeded, 92, 95, 96
time state, 153
time-deadlock, 182
time-extended state,
     see time state

time-
time-m
   initi
   integ
Timed Pe
Timed Petr
Timed Petri
   ration
transition-mar.
   124
tw-Petri net, 174

ultimo property, 1

zero-test, 39, 142, 18

Printed by Publishers' Graphics LLC
LMO131121.15.22.72

Index

$REIS_{\mathcal{Z}}$, <u>85</u>, 86
$RIS_{\mathcal{Z}}$, <u>52</u>, 71, 73
$RS_{\mathcal{Z}}$, 38
$R_{|\mathcal{P}|}$, 181
$R_{\mathcal{N}}$, 13
$R_{\mathcal{Z}}$, 38
$S(\mathcal{D})$, 140
$S(\mathcal{P})$, 174
$S(\mathcal{Z})$, 32
$\mathcal{CG}_{\mathcal{N}}$, 15
$\mathcal{IR}_{\mathcal{P}}$, 183
$\mathcal{M}ax(p)$, 98
$\mathcal{M}in(p)$, 98
$\mathcal{RG}_{\mathcal{D}}$, 145
$\mathcal{RG}_{\mathcal{N}}$, 14
$\mathcal{RG}_{\mathcal{Z}}$, 72, 77
$\mathcal{RG}_{\mathcal{Z}}^{w}$, 129
$\mathcal{RG}_{\mathcal{Z}}^{redu}$, 75, 87

AC net, <u>21</u>, 112

bag matrix, 156
BFC net, <u>112</u>, 113
blocking-free, <u>22</u>, 23, 94
boundedness
    in a Petri net, 13
    in a Time Petri net, 38
    in a Timed Petri net, 145
    in a tw-Petri net, 181

Computability, 17

computable
    CM-, 40
    PN-, 18
    Timed PN-, 148
    TPN-, 42
    Turing-, 17
conflict
    dynamic, <u>21</u>, 191
    static, <u>21</u>, 191

distance of time
    between p-markings, <u>121</u>, 130, 132
    between states, <u>119</u>, 129, 131

EFC net, <u>21</u>, 99
elapsing of time
    in a Time Petri net, 36
    in a Timed Petri net, 144
    in a tw Petri net, 177
enabled, 11

FC net, 21
firing
    in a Petri net, 11
    in a Time Petri net, 35
        modified, 78
    in a Timed Petri net, 144
    in a tw-Petri net, 176
firing sequence
    in a Petri net, 11
    in a Time Petri net, 38

in a Timed Petri net, 155
in a tw-Petri net, 177
function
    duration, <u>140</u>, 149
    interval, <u>32</u>, 33, 124, 126, 127
    number-theoretical, <u>6</u>, 39, 42, 148
    priority, 166

global reachability, 155
global step, 154

homogeneous, <u>21</u>, 99, 113

incidence matrix, 10, 24
    time-extended, 156

liveness
    in a Petri net, <u>22</u>, 23
    in a Time Petri net, <u>94</u>,
        96–98, 113
    in a Timed Petri net, 145
    in a tw-Petri net, 188

marked graph, 21
marking, <u>8</u>
    t^+, <u>10</u>
    t^-, <u>10</u>
    generalized, <u>14</u>
    initial, <u>8</u>, 9
    reachable, 10, 13, <u>13</u>
maximal step, 4, <u>141</u>, 142
    prioritized, 166
modified modulo, 176
multiset, 154

P-invariant, 25
Parikh matrix, 157
Parikh vector, <u>12</u>, 24, 118
Petri net, <u>7</u>

Petri net with retention time on places,
    see tw-Petri net
place-marking (p-marking), <u>33</u>, 34, 40,
    45, 92
    reachable, <u>38</u>, 51, 67, 71, 78, 85,
        86, 110
progress matrix, 157

reachability graph
    of a Petri net, <u>14</u>
    of a Time Petri net, 72, 75, <u>87</u>,
        191
    of a Timed Petri net, 145
    of a tw-Petri net, 183
ready to fire
    in a Time Petri net, 35
    in a tw-Petri net, 176
run, 37
    feasible, 37
    integer-, <u>52</u>, 67
    length of a, 118
    maximum, <u>119</u>, 121
    minimum, <u>119</u>, 121
    parametric, 46

skeleton
    of a Time Petri net, 32
    of a Timed Petri net, 140
    of a tw-Petri net, 174
state
    in a Time Petri net, <u>34</u>
        essential-, 85
        initial, 34, 124
        integer-, <u>51</u>, 67
        parametric, 46
        reachable, 38
    in a Timed Petri net, 141
state class, <u>49</u>

state equation
    in a Petri net, <u>12</u>, 151
    in a Timed Petri net, <u>158</u>,
        <u>162</u>

T-invariant, <u>24</u>, 118
time dimension, 152
Time Petri net, <u>32</u>
    finite, 71
    infinite, 78
    lazy, <u>92</u>, 93, 95, 97
    speeded, <u>92</u>, 95, 96
time state, 153
time-deadlock, 182
time-extended state,
       see time state

time-gap, 184
time-marking, 175
    initial, 175
    integer, 175
Timed Petri net, <u>140</u>
Timed Petri Nets with Priorities, 165
Timed Petri Nets with Variable Durations, 168
transition-marking (*t*-marking), <u>34</u>, 36, 124
tw-Petri net, <u>174</u>

ultimo property, <u>178</u>, 179

zero-test, 39, <u>142</u>, 186, 187